T0092024

Body Am I

Body Am I

The New Science of Self-Consciousness

Moheb Costandi

The MIT Press
Cambridge, Massachusetts
London, England

The MIT Press would like to thank the anonymous peer reviewers who provided comments on drafts of this book. The generous work of academic experts is essential for establishing the authority and quality of our publications. We acknowledge with gratitude the contributions of these otherwise uncredited readers.

This book was set in ITC Stone Serif Std and ITC Stone Sans Std by New Best-set Typesetters Ltd. Printed and bound in the United States of America.

Library of Congress Cataloging-in-Publication Data

Names: Costandi, Moheb, author.
Title: Body am I : the new science of self-consciousness / Moheb Costandi.
Description: Cambridge, Massachusetts : The MIT Press, [2022] | Includes bibliographical references and index.
Identifiers: LCCN 2021061882 (print) | LCCN 2021061883 (ebook) | ISBN 9780262046596 (hardcover) | ISBN 9780262368704 (epub) | ISBN 9780262368711 (pdf)
Subjects: LCSH: Self-consciousness (Awareness) | Cognition. | Neurosciences.
Classification: LCC BF311 .C697 2022 (print) | LCC BF311 (ebook) | DDC 153—dc23/eng/20220411
LC record available at https://lccn.loc.gov/2021061882
LC ebook record available at https://lccn.loc.gov/2021061883

10 9 8 7 6 5 4 3 2 1

For my children: Discover your selves, and have the courage to be who you are, not who others want you to be.

Contents

1 Introduction

> The awakened and knowing say: body am I entirely, and nothing more; and soul is only the name for something about the body.
>
> —Nietzsche, *Thus Spoke Zarathustra*

On Monday, September 17, 1810, Sir Joseph Banks, president of the Royal Society of London, joined other scientists, naturalists, and members of high society for a private viewing in the society's house of exhibition at 225 Piccadilly. The exhibit was a twenty-one-year-old Khoisan woman named Saartjie Baartman, who had arrived in England from the Cape of Good Hope several months earlier.

Baartman stood a little under 5 feet tall and, like other Khoisan women, had large breasts, protruding buttocks, and elongated vaginal lips (labia minora), said to resemble a turkey's wattle. She had been brought to Liverpool earlier that year by Alexander Dunlop, a British naval surgeon who used his position to traffic and trade museum specimens from the colonies. He had convinced Baartman that she could make a fortune exhibiting her body in Europe, and so it was that she traveled to England, along with various other exotic specimens Dunlop had collected.

There Baartman was subjected to medical observation and then exhibited as a freak show attraction, which members of the public could view at a cost of two shillings. The "Hottentot Venus," as she was called, was advertised as having "the kind of shape which is most admired by her countrymen"; on display, she wore a small, tight dress that made "her shapes above and the enormous size of her posterior parts . . . as visible as if [she] were naked," and she stood on a "stage two feet high, along which she was led by her keeper and exhibited like a wild beast, being obliged to walk, stand or sit as he ordered." Upon attending one of her shows, the theater manager and comedian Charles Matthews "found her surrounded by many persons, some *females*! One pinched her; one gentleman poked her with his cane; one *lady* employed her parasol to ascertain that all was, as she called it '*nattral*.' This inhuman baiting the poor creature bore with sullen indifference, except upon some provocation, when she seemed inclined to resent brutality. . . . On these occasions it took all the authority of the keeper to subdue her resentment."

This objectification continued long after Baartman's death. She was eventually sold to an animal trainer and showman named Réaux, who exhibited her in Paris for some eighteen months, before her death in December 1815, from what the French naturalist and zoologist Georges Cuvier diagnosed as "une maladie inflammatoire et éruptive." Upon her death, Baartman was autopsied by Cuvier, who made a plaster model of her brain and preserved her buttocks and vagina for museum display. He also published a detailed account of her anatomy, and used it as evidence for his scientific racism. Baartman's brain, skeleton and genitalia remained on display in the Musée de L'Homme until 1974. Twenty years later, after election as president of South Africa, Nelson Mandela requested that her remains be repatriated, and the French government capitulated in 2002.[1]

For her audiences, Baartman was a source of fascination and revulsion, who epitomized the ugliness, savagery, and primitive sexuality of the inferior races. For many today, however, the shameless treatment she received epitomizes the racist and exploitative nature of colonialism. But Baartman's tragic story also highlights our fascination with the form and appearance of the body. Now, as then, our societies are obsessed with the human body—perhaps even more so. People whose body is perceived to be beautiful are both celebrated and objectified, whereas those whose body is deemed to be ugly may be "body shamed," or worse. Partly in response to this, the popularity of the body positivity movement, which advocates the acceptance of everyone's body regardless of size, shape, or physical ability, has grown enormously in recent years.

Our obsessive fascination is reflected in the many, increasingly popular television shows about the body. In the United Kingdom, Channel 4's long-running medical reality show *Embarrassing Bodies*, in which members of the public consult celebrity doctors about a wide variety of medical conditions, has won multiple awards and spawned numerous spin-offs, including *Embarrassing Fat Bodies*, *Embarrassing Teenage Bodies*, and *Embarrassing*

Bodies: Live from the Clinic; in Australia, *Embarrassing Bodies Down Under;* in the Netherlands, *Dit is Mijn Lijf* (*This is My Body*); and in Ukraine, Я соромлюсь свого тіла (*I'm Ashamed of My Body*). The latest body-based television show in the United Kingdom is the dating game show *Naked Attraction*, where contestants appear on stage completely naked and gradually reveal their bodies from the feet up to another contestant, who chooses between them based solely on their physical appearance.

Images of the body are pervasive, and hundreds of millions of people around the world aspire to attain the "perfect" body, as portrayed—unrealistically—in the mass media. Women in particular have always felt enormous social pressure to look a particular way, but this pressure is now increasingly applied to girls of ever-younger age, as well as to boys and men. Our desire to improve our physical appearance fuels, globally, a men's grooming industry worth $50 billion per year, a health club industry that generates around $100 billion annually, a fashion retail market valued at more than $300 billion, and a cosmetics market worth over $500 billion. In 2019, more than 18 million cosmetic procedures were performed in the United States alone, including 299,715 breast augmentation procedures (the most popular cosmetic surgical procedure since 2006), 265,209 liposuction procedures, 211,005 eyelid surgeries, 207,284 nose reshapings, and 123,685 face lifts. The number of facial rejuvenation procedures performed has also increased every year since 2000. Of these, botox injection is by far the most popular, with 7.7 million injections performed in 2019, the highest annual number to date. Not surprisingly, more than 90 percent of these procedures were performed on women; worryingly, though, 223,000 were performed on teenagers.[2] The main factors that increase the likelihood of undergoing cosmetic surgery are low self-esteem, low life satisfaction, low self-rated attractiveness, and heavy television watching.[3]

According to Renee Engeln, a professor of psychology and director of the Body and Media Lab at Northwestern University, this cultural obsession with appearance amounts to nothing less than a society-wide psychological illness: "Beauty sickness is when you are so worried about how you look that it's taking your time, attention, and your emotional resources away from things that you would rather be spending those resources on."[4]

Following the development of functional neuroimaging technology in the early 1990s, neuroscientists began to investigate the "neural correlates of consciousness." These sophisticated new methods enabled them to identify the neurological substrates that support our conscious experiences. "Being conscious means that one is having an experience—the subjective, phenomenal 'what it is like' to see an image, hear a sound, think a thought or feel an emotion," begins a 2016 review by four prominent researchers. Early experiments typically involved scanning the brain to identify those areas and cell populations activated in response to specific "contents of consciousness," such as images of faces or other visual stimuli, as they drifted in and out of conscious awareness. Subsequently, researchers examined the brain mechanisms necessary for being conscious in a broader sense by comparing the brain activity associated with the "normal" conscious state to that seen during sleep, under general anesthesia, and in the various disorders of consciousness (the minimally conscious state, vegetative state, and coma).[5]

This approach identified the neural correlates of consciousness as a diffuse network of neural structures distributed throughout the cerebral cortex and brain stem, but it largely ignored half of the problem. Being conscious does not just mean having the subjective experience of events in the outside world. It also means having self-consciousness—that is, being aware of our self within our surroundings. As self-conscious creatures, we are aware of our self as *ourselves*. We think of our self as having a personality that is distinct from that of others, and we maintain a repository of autobiographical memories that gives our existence a sense of narrative or of purpose. But the core component of self-consciousness, which philosophers sometimes call the "minimal self," is bodily awareness, or our experience of our body.

Our body is central to our sense of identity and is often described as a canvas for self-expression. Throughout the ages, we humans have adorned our bodies with clothing and jewelry, decorated them with piercings and tattoos, and modified and mutilated them in an astonishing variety of ways. Some such practices are pathological; in the New Testament, for example, Mark 5:2,5 describes "a man with an unclean spirit," who "always, night and day . . . was in the mountains and in the tombs, crying out and cutting himself with stones," and Herodotus describes the Spartan leader Cleomenes, who became psychotic, took a knife, and, starting at his lower leg, "sliced his flesh into strips, working upward to his thighs, hips and sides until he reached his belly, which he chopped into mincemeat." Other practices, such as earlobe stretching by the Maasai peoples of Kenya or the

Jewish and Muslim traditions of circumcising male infants, are culturally or religiously sanctioned, while some newer practices—such as saline infusions, transdermal implants, tongue splitting, and flesh stapling—blur the distinction between body modification and self-mutilation.[6]

Historically, the mind and the brain were regarded as being separate from each other and made of completely different substances—the one mental, the other physical. Most modern neuroscientists agree that this is a false dichotomy—that the mind is made of matter and is a product of the brain. The old distinction between the mind and the brain has, however, been replaced by a new distinction between the brain and the body. Yet research into bodily awareness tells us that the brain and body are inextricably linked. The body is not just a canvas for self-expression or an interface between the brain and the outside world. Our brain controls our body, which acts as a conduit through which we perceive and act upon our environment, but our body in turn influences our perceptions, our thoughts, and our emotions. Yet most of us take our body for granted most of the time—in the words of William James, the father of modern psychology, we experience "the feeling of the same old body always there."[7]

Body Am I is about how our brain perceives our body, how that perception translates into our conscious experience of body, and how that experience contributes to our sense of self. The main idea presented within the book is that bodily awareness is fundamental to self-consciousness, and it is assumed that bodily awareness comprises two core components: body ownership, the sense that our body belongs to our self; and agency, the sense that we are in control of our body, our thoughts, and our actions. We know that body ownership and agency are related because they can influence each other. But we also know that they are distinct because certain neurological conditions disturb one sense but not the other: thus, for example, the rare condition commonly (but incorrectly) referred to as alien hand syndrome disrupts agency but not body ownership, such that those having the condition claim that their "alien hand" is under the control of some external force even though it belongs to them. By contrast, somatoparaphrenia disrupts body ownership but not agency, causing those who have it to claim that an affected limb belongs to somebody else, even though they retain full control of it.

One theme of *Body Am I* is that the brain generates maps and models of the body that guide how we perceive and use our body. Another theme is that bodily awareness is extremely malleable. The brain constructs these maps and models from fragments of sensory information it receives from the body, and repeatedly modifies and reconstructs them. Everyday experiences can alter these body maps and models in subtle ways; brain injury and disease can change them more dramatically, leading to a huge variety of bodily experiences that can profoundly change our sense of self.

The scientific investigation of bodily self-consciousness has a long history, with contributions from eminent nineteenth- and early twentieth-century neurologists, whose clinical observations provided valuable insights. One key figure in that history is Silas Weir Mitchell, who performed hundreds, or perhaps thousands, of limb amputations on the battlefields of the American Civil War. Mitchell recognized that the vast majority of amputees retain a ghostly impression of their amputated member and named the phenomenon "phantom limb." Another is the British neurologist Henry Head, whose treatment of patients with brain damage from the slums of East London linked bodily awareness to the parietal lobe. Head was one of the first to theorize the existence of a body model, which he called the "body schema," and the concept remains influential more than a century later. Beginning in the 1920s, the Canadian neurosurgeon Wilder Penfield went on to make profoundly important discoveries. Penfield developed a method of electrically stimulating the cerebral cortex in patients who remained fully conscious and, in the process, charted maps of sensation and movement on the surface of the human brain. Equally important, but lesser known, is the German neuropsychiatrist Paul Schilder, whose 1935 book *The Image and Appearance of the Human Body* introduced the concept of the "body image," which became highly influential in neurology and psychiatry and is now used widely in the humanities and social sciences. Schilder defined body image as "the picture of our own body which we have in our mind," and the term has become part of the popular lexicon, but very few people, even within neurology and psychology, recognize Schilder as its originator.

Schilder is arguably the greatest little-known physician and academic of the twentieth century. He conceptualized the body image as a multidimensional construct, with biological, psychological, and sociological components. The biological component consists of the body maps in the brain, and these are the basis of the psychological component of the body image, or the way we perceive, and think of, our bodies. But humans are social animals, and other people play a role in how we think of our bodies and our selves. Bodies interact with each another in numerous ways throughout life. A parent's loving caresses are crucial for the proper development of a newborn baby's body maps and sense of touch, and, later on, the child will learn behaviors by observing adults' bodies and imitating their actions. Mass media—and, increasingly, social media—play a large role in how we think about our bodies, causing many of us to experience dissatisfaction with them.

Head and Schilder were pioneers who laid the groundwork for modern bodily awareness research, which emerged as a field in its own right in the late 1990s, following the publication of a short letter in the scientific journal *Nature*, describing a simple illusion of bodily awareness in which experimental subjects could be tricked into thinking that a fake hand belonged to their self. The "rubber hand illusion" gave researchers an experimental model with which they could probe and manipulate the processes underlying bodily awareness. The field of research blossomed quickly, and today there are dozens of labs around the world devoted to understanding how we perceive our bodies, how these perceptions can be altered by experience, and how such processes relate to our thought processes, emotions, and sense of self.

Body Am I explores bodily awareness research and the new science of self-consciousness that has emerged from it. It traces the historical milestones from classical neurology and describes the latest cutting-edge research. The study of bodily awareness is a multidisciplinary endeavor that straddles neuroscience, psychiatry, and psychology, and findings from the research are providing fresh perspectives on a wide variety of conditions, including autism, eating disorders, and schizophrenia. It helps us to understand alien hands, phantom limbs, and other strange neurological conditions, such as body integrity identity disorder, which drives some individuals to amputate a perfectly healthy arm or leg, and it explains a whole range of other phenomena, from doppelgängers to ouija boards. The research is opening up new avenues of treatments for neurological and psychiatric conditions, and it is allowing for the development of next-generation prosthetic limbs, among other things. It is also forcing us to rethink the nature of consciousness—in our own species and others.

2 Bodily Awareness

Our perception of the world around us is constructed by our brain from limited information received by our sense organs. So, too, is our perception of our body. The brain perceives the body as another object in space, based on sensory information it receives, and this information gives rise to our conscious awareness of our body. The body is a very special object, however, because we experience it, not as an external object, but from a subjective perspective, as our *self*. That is to say, our sense of self-identity is closely linked to our body, and bodily awareness is a basic form of self-consciousness. Bodily awareness is malleable and easily distorted or disturbed. One particularly striking example of a disturbance in bodily awareness is phantom limb syndrome, in which an amputee vividly experiences a missing limb as still being attached to the amputee's body. Phantom limbs were described in the sixteenth century by the French barber-surgeon Ambroise Paré. Until then, few in the medical community took phantom limb syndrome seriously, but that eventually changed, with advances in medicine and military technology that dramatically reduced the mortality rate associated with major amputation. On the battlefields of the American Civil War, the neurologist Silas Weir Mitchell amputated thousands of limbs, and the higher survival rates of his patients finally enabled better observation and more detailed description of the phenomenon.

On September 19, 1863, George Dedlow awoke to find himself lying under a tree behind a Union field hospital, surrounded by wounded men, with doctors busying themselves at an improvised operating table made from two barrels and a plank of wood.

In April 1861, Dedlow had enlisted as an assistant surgeon in the Fifteenth Indiana Cavalry, but heavy casualties of the drawn-out war had thinned the ranks, so he had been promoted to first lieutenant in the Seventy-Ninth Indiana Volunteers and sent to the front line near Nashville. With supplies running low, Dedlow volunteered to go to a military outpost farther north, in order to obtain quinine and medicinal stimulants.

On his way to this outpost, he had been shot and captured by Confederate guerrillas. "A ball had passed from left to right through left biceps and directly through the right arm just below the shoulder, emerging behind," he later wrote. "The right hand and forearm were cold and perfectly insensible. I pinched them as well as I could, to test the amount of sensation remaining; but the hand might as well have been that of a dead man. I began to understand that the nerves had been wounded, and that the part was utterly powerless."

Dedlow's captors had placed him onto a horse-drawn cart and sent him off to a rebel hospital near Atlanta. The journey was bumpy and painful, yet reassuring: "The jolting was horrible, but within an hour I began to have in my dead right hand a strange burning, which was rather a relief to me. It increased as the sun rose and the day grew warm, until I felt as if the hand was caught and pinched in a red-hot vise."

With the long arm bone from shoulder to elbow badly broken and the nerves badly injured, Dedlow's right arm remained "red, shining, aching, burning . . . perpetually rasped with hot files" for weeks. It caused him, on one occasion, to faint from the pain and weakness, and when he awoke, the pain only grew worse; it became excruciating for more than a month, until at last the arm was amputated very near to the shoulder, and because the hospital's meager supply of ether and chloroform had run out, the operation was performed without anesthesia. "The pain felt was severe, but it was insignificant as compared with that of any other minute of the past six weeks."

Dedlow had been released from captivity in early April 1863, in exchange for a Confederate prisoner and, after thirty days' leave, had returned to his regiment as a captain. In September of that year, the regiment had joined in the Battle of Chickamauga, and Dedlow had been injured again when a battery opened fire from the left as he climbed a hill toward a rebel-held fortress.

And so it was that Dedlow woke up under a tree, with one hospital steward raising his head to pour brandy and water into his mouth, and another cutting loose his pantaloons. When he asked one of the stewards where he had been injured, he was told he had been wounded in both thighs, and that there was little the doctors could do. A moment later, the steward placed a towel over his mouth, and Dedlow smelled the familiar odor of chloroform. "The trees began to move around from left to right, faster and faster; then a universal grayness came before me."

Dedlow remembered nothing more until regaining consciousness in a hospital tent. Feeling a sharp cramp in his left leg, he tried to rub it with his single remaining arm but found himself too weak. When, however, he hailed an attendant to do it for him, telling him that he now felt pain in both of his legs, the attendant looked at him in surprise and threw the covers off. Dedlow looked down and, to his horror, saw that both of his legs had been amputated.[1]

The American Civil War was a brutal conflict, in which some 600,000 men died and some 500,000 more were maimed, mutilated, or otherwise disfigured. This high casualty rate was due in large part to the introduction and widespread use of a newly developed bullet called the Minié ball. Named after its codeveloper Claude-Étienne Minié, the Minié ball was actually cylindrical with a cone-shaped hollow cavity and three grease-filled grooves around its base; it was fired from a specially made weapon with grooves rifled on the inner surface of the barrel. Like the smooth-bore musket balls that preceded them, Minié balls were muzzle loaded but much smaller than the gun bore, thus far easier to load in combat. They were designed to expand under pressure when fired, which increased their muzzle velocity, and to spin as they moved through the barrel, increasing their aerodynamic stability. This allowed for both greater accuracy and far longer range. And because they were made from soft lead, Minié balls flattened out or splintered upon impact, shattering bones, ripping through flesh and other soft tissue, and carrying pieces of skin and clothing into the wound.

Invented in 1849, Minié balls were used widely in the American Civil War by both Union and Confederate soldiers. As a result, untold numbers of soldiers not killed outright by these more deadly bullets suffered grievous gunshot wounds from the bullets' fragmentation. Amputation therefore became a "signature" feature of that conflict.

"In steady hands, frightful wounds are produced by the Minié ball, which require all the resources of surgery to manage successfully," wrote

Julian J. Chisolm in the 1864 edition of the Confederate *Manual of Military Surgery*. Another military surgeon in the Twenty-First Kentucky Infantry wrote to his wife, in a letter dated June 29, 1864, that "my hands are constantly steaped in blood. I have amputated limbs until it makes my heart ache to see a poor fellow coming in the ambulance."[2]

Physicians of the time favored minimal intervention, in order to preserve the integrity of the patient's body. When treating a limb injury, splinting and surgical excision of the damaged bone were preferred over amputation, which was only to be performed when absolutely necessary. The reality of battlefield surgery was quite different, however. William Williams Keen, a surgeon from Philadelphia who served in the US Army during the Civil War, later described the "sweet . . . mouse-like" smell of gangrene that filled the air of the field hospital, recalling that he and his colleagues "operated in old blood-stained and often pus-stained coats . . . with undisinfected hands . . . [reusing] marine sponges used in prior pus cases and only washed in tap water." Another observer describes surgical treatment of soldiers wounded in the Battle of Gettysburg: The surgeon "snatched the knife from between his teeth, where it had been while his hands were busy, wiped it rapidly once or twice across his blood-stained apron, and the cutting began. The operation accomplished, the surgeon would look around with a deep sigh, and then [call]: 'Next!'"

Splinting and excising bone ran the risk of infection, deformation, and stiffening or fixation of the limb's joint ("ankylosis"), which could render the limb useless. The US Sanitary Commission therefore recommended that a limb be amputated if it had sustained a compound fracture or was badly lacerated. And surgeon Chisolm advised amputation (1) when a limb or joint had been crushed; (2) when the major blood vessels or nerves had been torn, even when the bone itself was unharmed; (3) when the soft tissue was severely cut or the skin extensively destroyed; or (4) when gangrene had set in. Far from being an exception to the rule, amputation was performed routinely, becoming so common that Confederate nurse Kate Cumming reported it to be "scarcely noticed" in the Georgia field hospitals at which she worked. In all, an estimated 30,000 to 60,000 limb amputations were performed during the four-year war, more than during any other war in which the United States has been involved.

George Dedlow's amputations were not among them, however, for "The Case of George Dedlow" is a fictional account of phantom limbs, written by the neurologist Silas Weir Mitchell and published anonymously in the July 1866 issue of *The Atlantic Monthly*.

Mitchell had graduated from Jefferson Medical College in 1850. When the Civil War broke out in April 1861, he enlisted as a contract surgeon and was assigned to the Filbert Street Military Hospital in Philadelphia, where he treated wounded soldiers, and "began to be interested in injuries and wounds of nerves, about which little was known." The following year, US Army Surgeon General William A. Hammond established the 400-bed Turner's Lane Military Hospital for Nervous Diseases, and Mitchell became its director.

Mitchell worked with his colleagues William Keen and George Morehouse, treating soldiers who were "worn out from fever, dysentery and long marches," and who had suffered "every conceivable form of nerve injury," including injuries from gunshots, shells, and saber cuts. Following the Battle of Gettysburg in early July 1863, the hospital at Turner's Lane filled with wounded soldiers; so, too, did Philadelphia's South Street Hospital, which became so closely associated with the treatment of amputees that it was called the "Stump Hospital."[3]

Mitchell, Keen, and Morehouse had had little knowledge of neurology and little or no experience in treating neurological patients before the war. But during the conflict, and for years after it had ended, they treated hundreds of soldiers with nerve injuries, examined them thoroughly, and documented their observations in thousands of pages of comprehensive notes. "Several nights each week we worked at note-taking often as late as twelve or one o'clock in the morning," Keen recalled, "and when we got through, we walked home a couple of miles talking over our cases. . . . The opportunity was unique and we knew it. The cases were of amazing interest."

Their long-lasting collaboration was extremely fruitful and led to publication of several influential books, most notably, *Gunshot Wounds and Other Injuries of Nerves*, which contains the earliest description of what we now call "shell shock," published in 1864, and *Injuries of Nerves and Their Consequences*, published in 1872. By this time, Mitchell was an eminent medical consultant on "nervous diseases" with a large practice; he would go on to become the first elected president of the American Neurological Society. His work at the Turner's Lane Hospital effectively laid the foundations of American neurology, and he is now considered to be the founding father of the field.[4]

Mitchell noted that the vast majority of his amputee patients experienced vivid sensations that their missing limb was still attached to their body. "Nearly every man who loses a limb carries about with him a constant or inconstant phantom of the missing member," he wrote in *Injuries of Nerves and Their Consequences*, "a sensory ghost of that much of himself, and sometimes a most inconvenient presence, faintly felt at times but ready to be called to his perception by a blow, a touch or a change of wind. Very many have a constant sense of the existence of the limb, a consciousness even more than exists for the remaining member." Mitchell also noted that these sensory hallucinations could be so vivid that they caused "absurd mishaps": "The sufferer who was lost a leg gets up in the night to walk, or he tries to rub or scratch it. One of my cases attempted, when riding, to pick up his bridle with the lost hand . . . and was reminded of his mistake by being thrown. Another person, for nearly a year, tried at every meal to pick up his fork, and was so disturbed emotionally at the result as frequently to be nauseated, or even to vomit."[5]

This puzzling phenomenon had been observed centuries earlier by the French barber-surgeon Ambroise Paré (1510–1590). Considered to be one of the fathers of early modern surgery, Paré revolutionized battlefield medicine and made major contributions to the treatment of amputees. At the time, amputation wounds were first cleaned with a solution of boiling oil and then seared with a red-hot iron. This often failed to stem the bleeding, however, and patients just as often died of shock. But one day in 1537, Paré ran out of his boiling oil solution and applied instead a soothing balm of egg yolks, rose oil, and turpentine. He also applied ligatures to large blood vessels that were severed during the operation—a procedure first used by the Greek physician Galen in the second century CE—and tourniquets (whose use Galen opposed) above the amputation site to stanch the bleeding.

Before the introduction of anesthetics and antiseptics, amputation was one of the most challenging operations a surgeon could perform. It had to be done quickly, with a single, circular cut through skin, muscle, bone, and tendon. A skilled barber-surgeon could perform the procedure in under five minutes, but the mortality rate was high: between 50 and 80 percent of all amputation patients died on the operating table.

Paré reportedly amputated some 200 limbs per day on the battlefields of Italy, but, thanks to his innovations, more of his patients survived the operation and lived to tell strange tales of their experiences as amputees. "Verily it is a thing wondrous, strange and prodigious," Paré wrote: "[T]he patients who have many months after the cutting away of the leg grievously complained that they yet felt exceeding great pain of that leg cut off."[6]

As the French mathematician and philosopher René Descartes (1596–1650), born six years after Paré died, also noted: surgeons "know that those whose limbs have recently been amputated often think they still feel pain in the parts they no longer possess." He described the case of a girl who "had her whole arm amputated because of creeping gangrene" and who complained for weeks after the operation "of feeling various pains in her fingers, wrist, and forearm."[7] But it was Silas Mitchell who first referred to this phenomenon as "phantom limb"—the name it goes by today—and who provided the first full medical description of it, in "The Case of George Dedlow."

Phantom limb sensations were poorly understood, and many of Mitchell's contemporaries likened them to paranormal phenomena, an idea that the term "phantom limb" may have contributed to. Thus it is often said that Mitchell published his observations in the form of a short story for fear of ridicule by the medical and scientific communities. Indeed, at the start of his account, Dedlow admits that "notes of my own case have been declined . . . by every medical journal to which I have offered them."

According to another account, Mitchell never intended to publish the story. One evening, in a discussion about limb amputation with his school friend Henry Wharton, he mentioned hearing about a man who had lost both arms and both legs in 1864 during the Battle of Mobile Bay. This inspired him to write the story; later, he saw the name "Dedlow" on a jeweler's sign, and thought it a suitable one for his protagonist. After writing the story, he gave it to the physician Caspar Wistar's wife, who passed it to her father, Dr. Furness, who then sent it on to Reverend Edward E. Hale, editor of *The Atlantic Monthly*. In due course, much to his "surprise and amusement," Mitchell received the page proofs, together with a check for $80.[8]

"The Case of George Dedlow" conveyed Mitchell's extensive experience of treating amputees at Turner's Lane Hospital, combining accurate and detailed medical information about Civil War injuries with a realistic first-person narrative. The story is of historical interest, both as one of the earliest attempts to describe medicine and surgery from the patient's point of view and as a vivid account of the high cost of armed conflict. Although the title character was apparently based on the experience of one very unfortunate individual, Dedlow can be thought of as a composite figure that represents the hundreds of thousands of soldiers who were maimed and mutilated in the Civil War.[9]

Through his protagonist, Mitchell speculated about the origins and causes of phantom sensations and pain, in terms of sensory impressions transmitted from the body to the brain. "The knowledge we possess of any [body] part is made up of the numberless impressions from without which affect its sensitive surfaces," he wrote, "and which are transmitted through its nerves to the spinal nerve-cells, and through them, again, to the brain. We are thus kept endlessly informed as to the existence of parts, because the impressions which reach the brain are, by law of our being, referred by us to the part from which they come."

Mitchell likened the nerves and their functioning to a wire bell pull. "You may pull it at any part of its course, and thus ring the bell as well as if you pulled at the end of the wire. . . . [I]n any case, the intelligent servant will refer the pull to the front door, and obey it accordingly." Thus "when the [body] part is cut off, the nerve-trunks which led to it and from it, remaining capable of being impressed by irritations, are made to convey to the brain from the stump impressions which are as usual referred by the brain to the lost parts, to which these nerve-threads belonged." Mitchell further argued that sensory impressions continue to be sent to the brain until the amputation stump has healed and that, in cases of incomplete recovery, these impressions cause constant irritation or 'burning neuralgia.'"[10]

Dedlow states that his account was declined by the medical journals "because many of the medical facts which they record are not altogether new, and because the psychical deductions to which they have led me are not in themselves of medical interest." Several years later, Mitchell explained that he had taken "advantage of the freedom accorded to a writer of fiction" and never imagined that what he called his "humorous sketch, with its absurd conclusion, would for a moment mislead anyone."

Evidently, though, Mitchell underestimated the story's significance. At the time it was published, *The Atlantic Monthly* was the most widely read magazine in the United States, and "The Case of George Dedlow" became immensely popular. Despite being a work of fiction, its narrative was so realistic that readers took the story as fact. They assumed that it was an autobiographical account, and that Dedlow was a patient being treated at Philadelphia's South Street Hospital; indeed, many wrote letters to Dedlow care of the hospital, and some tried to visit him there. Some even raised money for his care.[11]

Today Mitchell is best known for his accurate description of phantom limbs. His experience of treating amputees left a lasting impression on him, however. Indeed, on his deathbed in January 1914, he was tormented by visions of maimed soldiers: his "wandering mind returned to those scenes," and his "wandering talk was of mutilation and bullets."[12]

For centuries after Paré's early description of them, phantom limbs were thought of as figments of the amputee's imagination. One British neurologist, writing in the early 1940s, expressed his dismay that, even "in these days of advanced physiological knowledge," medical men were "beyond reason," refusing "to believe that phantom limbs are anything more than psychological abnormalities." "[I]t would not be surprising if the unfortunate patient," he continued, "was regarded [by such men] as an obstinate, lying fellow or even possessed of the devil," and any amputee who experienced a phantom limb "must have often distrusted the reality of his own sensations" and would likely have avoided discussion of it for "dread of the unusual, of disbelief, or even the accusation of insanity."[13]

By explaining phantom limbs in physiological terms, Mitchell brought them into the realms of medicine. He clearly believed that phantom limb sensations were real, but that they were conjured up—at least in part—by the brain. And he described certain of their characteristics that have been recognized by neurologists ever since.

Phantom limb sensations in amputees seem to be the rule rather than the exception, as they are experienced by almost everyone who loses a limb. During World War II, a pair of British neurosurgeons studied 300 German prisoners of war with "major" amputations (an arm or a leg), and many more with "minor" ones (fingers or toes), and noted that "only about 2% of them asserted they never felt a phantom."[14]

In the vast majority of cases, amputees report that their phantom limb has a well-defined shape, adopts a certain position, is capable of voluntary, natural movement, and usually persists for years, sometimes decades, after amputation. And so we have, as reported in the journal *Brain* in 1941, the case of a forty-eight-year-old man, "a Methodist parson of somewhat rigid and unimaginative outlook," who had his right leg amputated just below the knee: "The phantom [foot] closely resembled in many ways his real foot. It was correctly placed in relation to his stump with which it moved, and, on voluntary effort, he could flex and extend the phantom foot and toes." Indeed, this patient's phantom felt so natural that "he was apt to forget that his foot was not real and attempted to use it in the ordinary way." He was "accustomed . . . to sit with his right amputated leg crossed over his left" and, "if he had occasion to get up suddenly from his chair, he often stepped off with his phantom foot." He had lost his leg at fourteen years of age but had remained aware of his phantom foot ever since and up to his neurological examination thirty-four years later. Then there is Tom Sorenson, who lost his left arm just above the elbow in a car accident, but remained aware of its ghostly presence afterward. "He could wiggle his 'fingers,' [and] 'reach out' and 'grab' objects that were within arm's reach." His phantom arm was so vivid that it "seemed to be able to do anything that the real arm would have done automatically, such as warding off blows, breaking falls or patting his little brother on the back." Sorenson had been left-handed before the accident—which happened before cellular phones became ubiquitous—and so he often found that his phantom arm "would reach out for the receiver whenever the telephone rang."[15]

Amputees sometimes describe their phantom limb in a positive light. Many of the amputees studied in the German prisoner-of-war camps and hospitals reported feeling "a pleasant sensation, usually described as tingling, or numbness, which is not painful and, indeed, is often reported as 'quite pleasant.'"[16] But at least half of amputees report feeling painful sensations, and in about one-tenth of them, this phantom pain is severe. These amputees variously describe their pain as a stabbing, throbbing, burning, or cramping sensation; their drawings provide valuable insights into how their phantom limb feels to them.[17] When they are asked to draw their phantom or to guide an artist to do so, the resulting images contain striking visual metaphors illustrating their pain—they show fingernails digging into the palm of the hand and drawing blood, a leg snapped in half with broken bones protruding from flaps of skin, and a hand or foot with nine-inch nails driven through it or crushed in a vise or under a lawn roller.[18] And although amputees often say that their phantom limb is capable of voluntary movement, pain is sometimes associated with the feeling that the phantom limb is "frozen" in the position the real limb had been in immediately before it was removed. Thus we have the case of "a soldier whose right arm was blown off by the premature explosion of a bomb, which he had been holding in his hand," who subsequently "felt as if his painful phantom hand were still grasping the bomb and he could not alter its position,"[19] and the case of a man who once had a sliver of wood caught under his fingernail and who, following amputation of his arm, reported sensing the splinter under the nail of his phantom finger.[20]

Amputees often say that their phantom limb changes with time, gradually retracting so that the digits get progressively closer to the residual stump. This is referred to as "telescoping," and is most apparent following a high amputation near the shoulder or pelvis. The time period over which this telescoping occurs varies widely from one individual to the next. In some amputees, the phantom digits might reach the stump within a month or two after the limb is removed; in others, the process might take several years. Usually, the length of the phantom limb decreases first, resulting in what many upper-limb amputees describe as "a baby arm with a hand of normal size." Phantom limbs usually "retain their normal size for some time . . . after the weaker parts have faded out of consciousness . . . before becoming gradually smaller."[21]

Phantom sensations and phantom pain are not restricted to amputated limbs. In a short footnote in his *Injuries of the Nerves and their Consequences*, Mitchell mentions phantom breasts and one case of a phantom penis, described to him by a US Navy physician: "These facts are not confined to lost limbs or parts of limbs. The amputated breast is often felt as if present, and the lost penis is subject to [phantom] erections, of which Dr. Rauschenberger, U.S.N., has related to me a remarkable example."[22]

The first detailed case report of phantom breasts was published in 1955 by surgeons at the University of Pennsylvania. Interviewing at varying times after surgery fifty women who had undergone mastectomies for breast cancer, they asked them: "Have you ever felt the presence of that [amputated] breast?" Eleven of them indicated that they had. One of the patients, a sixty-six-year-old woman interviewed eight months after the operation, told them, "It just feels like it is there sometimes. . . . I feel as though there is a tingling and slight itching in the nipple," adding that she experienced these sensations once or twice a week, and that they lasted for "just a couple of minutes." A fifty-six-year-old woman with eleven children, interviewed two-and-a-half years after her operation, said, "I felt like the nipple was full like it would be when it was ready to nurse . . . and I grabbed myself there." A third woman told them that right after the operation she "began to feel the heaving feeling of the breast or thought [it] itched me—and I'd reach over to scratch it," and another patient reported feeling that "the breast is shaking and moving with me when I walk." Only two of the patients reported phantom pain sensations, but neither said the pain was severe.

The researchers noted that the onset of symptoms varied widely. Some of the patients began to experience the phantom sensations or pain immediately after the operation; one did not experience these symptoms until more than two years later; but, for most patients, the symptoms began after the excision had healed. The duration of the symptoms also varied. One of the patients felt the phantom breast just once since her operation, whereas another felt it once or twice a week for four years.[23] A 2007 study by Dutch clinicians found a similar incidence, with fourteen out of eighty-two breast cancer patients (17 percent) experiencing phantom breast sensations two years after mastectomy. This percentage remained stable over time, whereas the incidence of those who experienced phantom breast pain was reduced from 7 to 1 percent over the same two-year period.[24]

Reports of phantom penises are far less frequent, and although the earliest description of them is usually attributed to Mitchell, the phenomenon was actually documented seventy-four years earlier by the prominent Scottish anatomist and surgeon John Hunter. A footnote in Hunter's 1792 book, *Observations on Certain Parts of the Animal Oeconomy*, describes two cases of phantom penis sensations, including the case of "a serjeant of marines who had lost the glans [head], and the greater body of the penis," who "declared, that upon rubbing the end of the stump, it gave him exactly the sensation which friction upon the glans [had] produced, and was followed by an emission of the semen."

In the best-known phantom penis case report, published in 1950, the Boston surgeon A. Price Heusner describes an older man whose penis was "accidentally traumatized and amputated," after which he "was intermittently aware of a painless but always erect penile ghost whose appearance was neither provoked nor provokable by sexual phantasies." These sensations were so vivid that the man had to look under his clothes to be sure that his penis was in fact missing.[25]

More recently, C. M. Fisher, a neurologist at the Massachusetts General Hospital in Boston, reported several other cases. One of these was the case of a successful businessman who developed a painful sore on the glans, which a biopsy revealed to be a carcinoma. The man underwent total penis amputation, at forty-four years of age. Twenty years later, while being treated in the hospital for a minor stroke, he casually mentioned that, ever since the amputation, he had regularly experienced phantom erections when sexually aroused, by, for example, "seeing a pretty young woman." His phantom erection "seemed to be of normal size, configuration and alignment," Fisher reported, and, just as it had to Heusner's patient, the phantom erection felt so real to him that he was "periodically obliged to check on the situation, tactually and visually." To the patient's surprise, his phantom erections always involved "the exact reproduction . . . of the original painful sore, at the same site on the glans, accompanied by the same type and severity of pain as before the operation."[26] Perhaps even more bizarrely, there are also reports of people experiencing phantom eyes, noses, and teeth, and one 1998 study revealed that more than two-thirds of patients who underwent amputation of the rectum later experienced phantom rectal sensations, including hemorrhoid-like pains, and even the presence of phantom feces and gas in the missing body part.[27]

Mitchell recognized phantom limbs as disturbances of bodily awareness. He understood that a missing limb remains in an amputee's conscious awareness long after it has been removed from their body, and that losing a limb altered his patients' sense of self-identity. In his account, Dedlow's last remaining limb is amputated after becoming gangrenous, and he remarks: "I found to my horror that at times I was less conscious of myself, of my own existence, than used to be the case." With all four limbs missing, he had now become "a useless torso, more like some strange larval creature than anything of human shape," and this caused "a deficiency in the egoistic sentiment of individuality," reducing him to a mere "fraction of a man." In the end, Dedlow attends a séance and is, for a moment, spiritually reunited with his limbs, which had been kept in storage for nine months, preserved "in the strongest alcohol" at the US Army Medical Museum in Washington, DC. "I was re-individualized, so to speak."

William James, the father of modern psychology, also recognized that the body is a core component of self-consciousness. His influential textbook *The Principles of Psychology*, published several decades after Mitchell's short story, includes a chapter titled "The Consciousness of Self," in which James tries to distinguish between what we call "me" and what we call "mine."

"[T]he line is difficult to draw," he states. "We feel and act about certain things that are ours very much as we feel and act about ourselves. Our fame, our children, the work of our hands, may be as dear to us as our bodies are, and arouse the same feelings and the same acts of reprisal if attacked. And our bodies themselves, are they simply ours, or are they *us*?" For James, the life of the self is divided into three constituents: the spiritual self, or "a man's inner or subjective being"; the social self, or "the recognition which he gets from his mates"; and the material self, of which "the body is the innermost part."[28] Modern philosophers and neuroscientists think in similar terms. They regard the body, and our experience of it, as the "minimal self," onto which other aspects of self-consciousness, such as memory and personality—sometimes referred to as the "extended self"—are superimposed.

Phantom limbs and other disturbances of bodily awareness have mystified investigators for millennia and, as the word "phantom" suggests, were often attributed to paranormal phenomena. Over the past century, some of the biggest names in neurology, neurosurgery, and psychiatry have contributed to our understanding of the neural basis of bodily awareness. Through clinical observations, laboratory study, and self-experimentation, these pioneers provided valuable insights and laid the groundwork for modern research into the subject. In the past twenty years, the neuroscience of bodily awareness has flourished as a field of research in and of itself, and today there are dozens of labs around the world devoted to studying how the brain perceives and interprets the body and to explaining both the various ways the mechanisms of bodily awareness can be disturbed and the myriad bizarre symptoms and behaviors that can occur when they are disturbed. These recent advances lead us to a new understanding of self-consciousness, of how the brain and body cooperate to generate our sense of self within the world around us.

Bodily awareness comprises two separate but related core components—body ownership, the sense that our body belongs to our self, and agency, the sense that we are in control of our body, our thoughts, and our actions. We know these are distinct from each other because one core component can be disturbed while the other remains unaffected. For example, people with a condition called somatoparaphrenia, which can occur after stroke damage to certain parts of the brain, firmly believe that one of their arms or legs does not belong to them, but do not deny that they are in control of the limb. Conversely, those with what is commonly called alien hand syndrome state that some external force is controlling the movements of their arm or hand, but do not deny that the limb belongs to them. Thus, in somatoparaphrenia, the sense of body ownership is disturbed, but the sense of agency remains unaffected, whereas, in alien hand syndrome, it is agency that is disturbed and body ownership that remains unaffected. In the case of phantom limbs, an amputee remains aware of a limb that is no longer attached to their body—thus retaining a sense of ownership over the missing body part.

The questions that bodily awareness researchers concern themselves with may seem trivial, such as "How do I know my body is mine?" and "How do I know I am in control of my body?" The brain mechanisms underlying the feelings behind these questions are, however, extraordinarily complex. Our senses of bodily awareness depend largely on continuous streams of sensory information from our body and the outside world, which enter the brain and are combined and interpreted by it, to give rise to our sense of bodily self-consciousness. The bodily self is, therefore, something like a jigsaw puzzle, constructed anew minute by minute within the brain from the fragments of information it receives from the senses. Most of the time, there is a unity in the way we perceive our body and ourself because the sensory fragments are pieced together accurately enough so that our perceptions appear seamless. Occasionally, though, one or other of the sensory streams is interrupted, or one of the mechanisms by which they are combined or interpreted goes awry, and it is then that the complex fragmentary nature of the bodily self reveals itself.

The new science of self-consciousness provides fresh insights into how the senses of bodily awareness are constructed. It provides a new understanding of a wide variety of neurological and psychiatric conditions, including rare disorders such as alien hand syndrome and somatoparaphrenia and of more common ones such as autism, eating disorders, and schizophrenia. And it also opens up new avenues of treatment for an equally diverse array of ailments and diseases—phantom pain being just one example.

The American Civil War vastly increased the demand for artificial limbs, and although many different models of artificial limbs were already available, the conflict drove advances in prosthetic technology and led to major growth of the prosthetics industry. For fifteen years preceding the war, official records show a total of 34 patents for artificial limbs, crutches, and "invalid chairs," compared to 76 patents for artificial legs, 19 for crutches, 8 for invalid chairs, and 18 for artificial arms, forearms, and hands for the twelve years from 1861 to 1873.

Inventors and manufacturers created devices that served both to help restore their users' day-to-day functioning and to hide or mask their injuries. Most of these early prostheses were cumbersome, unsightly, and uncomfortable, but some later innovations were much less so, including prostheses with combination knives and forks for upper-limb amputees, and a few devices of increasingly realistic feel and appearance. In her memoirs, union nurse Adelaide Smith writes that "it was surprising how many were well fitted with the limbs, and that they could walk so well that only a slight limp betrayed them," noting that arm amputees "with neatly gloved hands, which they could sometimes use quite well, were seldom observed in passing," and even describes one patient with a government-issue peg leg which "fitted so well that he could jump off a moving car."[29]

Limb amputation was the most common surgical procedure performed during the Civil War, and loss of limbs is also a signature injury of the post-9/11 conflicts in Afghanistan and Iraq. Accidental detonation of unexploded cluster bomblets in Afghanistan and the widespread use of improvised explosive devices (IEDs) in Iraq have led to the traumatic loss of one or more limbs in several thousand American and British military personnel since these wars began—and, of course, to similar injuries in unknown numbers of civilians.[30]

Research into the mechanisms of bodily awareness has provided a treatment that can alleviate the phantom pain sensations felt by a large proportion of amputees, and it is paving the way for the next generation of prosthetic limbs. These advanced prostheses will not only look like real limbs but will also be fully integrated into their users' nervous systems. This will make a prosthetic limb feel like a part of a user's body, rather than a cumbersome extension of it, and will give the user an unprecedented level of control over the new body part.

3 Body Ownership

As we noted before, one of the two core components of bodily awareness is body ownership—the sense that our body belongs to our self, that it is distinct from the bodies and the selves of others and from other objects in the world. In general terms, phantom limbs can be thought of as a disturbance of body ownership, in which amputees retain some degree of ownership over their missing limb. Various other disturbances of body ownership have been observed in the clinic. Patients with somatoparaphrenia, for example, deny ownership of a limb, firmly believing that it belongs to somebody else, although they retain control of it, as do patients with body integrity identity disorder (BIID), whose firm belief, however, develops into a burning desire to have that limb removed. The sense of body ownership can be easily manipulated in the laboratory. A simple experiment published in 1998 showed that study subjects could be readily induced to take ownership of a fake hand. In the "rubber hand illusion," now widely used and replicated thousands of times since 1998, subjects are manipulated into accepting a rubber hand as their own, enabling researchers to investigate the brain mechanisms underlying body ownership. The original rubber hand illusion has been modified in numerous ways, to create, for example, a "body swap illusion," that provides further insights into the sense of ownership.

Robert Vickers had always felt that there was something wrong with his left leg, even though the limb was perfectly healthy—it moved and felt as it should, and it was not deformed in any way. Yet, for some unknown reason, Vickers felt that his left leg was superfluous, that it just did not "belong" to him, and that he should have been born without it. He recalls having such thoughts from as early as five years of age. Growing up, he was unable to share these thoughts even with those who were closest to him, which caused him a great deal of anxiety. As a teenager, he became depressed and eventually attempted to end his life.

Vickers felt incomplete with the left leg attached to his body; indeed, he came to believe that the only way he could be "whole" was to have it removed. He would often pretend to be an amputee by tying the leg back with a belt, and he tried to damage the limb irreparably in order to force its surgical amputation. On one occasion, he constricted its blood flow with a tourniquet, and on another, he placed it beneath his jacked-up car in an attempt to crush it. But the leg proved to be tougher than he imagined, and his attempts to remove it were fruitless.

By the time he was thirty years old, Vickers had decided that he needed to get rid of the unwanted limb once and for all. So, one morning, he bought a large quantity of dry ice and submerged the leg into it. He then rang his wife and asked her to take him to the hospital, hoping that he would wake up the next day to find that his leg had been removed. Unfortunately for Vickers, the doctors managed to save the limb. They thought he was insane and gave him tranquillizers, antipsychotics, and electroconvulsive therapy in an attempt to rid him of his, to them, senseless desire. Although Vickers had long considered himself to be mad, too, he finally came to realize that having the limb surgically removed was the only thing he could do to be happy. Eventually, he met a psychiatrist who had treated several other similar patients, and who arranged for his leg to be amputated. His only regret was that he hadn't done it sooner.[1]

Vickers has a rare and mysterious condition now referred to as body integrity identity disorder. The first medical description of the condition appeared in 1977, in a short paper describing two cases of "self-demand amputation." One of the patients reported that he had recurring sexual fantasies about being an amputee, and that he often masturbated to photographs of female amputees. "Since my thirteenth year," he told the researchers, "my conscious life has been absorbed . . . in a bizarre and prepotent obsessive wish, need, desire to have my leg amputated above the knee." He had approached several doctors about his condition, but he was repeatedly refused a surgical amputation, and eventually began to imagine various accidents to injure his leg. One day, he took matters into his own hands—he hammered a tapered stainless-steel stylus into his left shin bone, and then attempted to infect the wound with facial acne pus mixed with nasal and anal mucus. His leg began to show signs of serious infection, and he admitted himself to the hospital, telling doctors that he had wounded the limb in an occupational accident. The infection cleared up quickly, however, and he was discharged. Afterward, he made many more attempts to injure his unwanted leg. He would regularly tie a tourniquet around his left thigh, using ice and injections of local anesthetic to numb the area, stopping only when the pain became unbearable. He then began hammering the stylus into his shin bone on a daily basis, moving closer and closer to the knee with each attempt. In follow-up interviews, the patient reported that he had contacted several psychiatrists and had considered various therapies, including LSD therapy, but it is unknown if he actually undertook them. As of the final follow-up, he still had not resolved his problem and reported being severely depressed.

BIID has been hiding in plain sight for centuries and has always been associated with sex. In 1785, the French anatomist and surgeon Jean-Joseph Sue describes an Englishman who was in love with an amputee and wanted to become an amputee himself. He offered a French surgeon 100 guineas to amputate his healthy leg; the surgeon refused because he did not have the right equipment but then performed the operation at gunpoint. The surgeon later received a payment for the amputation, along with a letter stating, "You have made me the happiest of all men by taking away from me a limb which put an invincible obstacle to my happiness."[2]

In the late nineteenth century, the Austro-German psychiatrist Richard von Krafft-Ebing published *Psychopathia Sexualis*, an encyclopedia of paraphilias, or deviant sexual practices, including bestiality, masochism, necrophilia, and sadism, along with a wide variety of other fetishes. Later editions of the encyclopedia, which is now considered to be a foundational text for the science of sexology, also include reports of three individuals who had what appears in hindsight to be BIID. "Even bodily defects become fetishes," von Krafft-Ebing wrote, before describing a twenty-eight-year-old factory engineer who "complained of a peculiar mania, which caused him to doubt his sanity":

> Since his 17th year he became sexually excited at the sight of physical defects in women, especially lameness and disfigured feet. Normal women had no attraction for him. If a woman, however, was afflicted with lameness or with contorted or disfigured feet, she exercised a powerful sensual influence over him, no matter

> whether she was otherwise pretty or ugly. In his dreams . . . the forms of halting women were ever before him. At times he could not resist the temptation to imitate their gait, which caused vehement orgasm with lustful ejaculation. . . . He thought it would cause him intense pleasure to mate with a lame woman. At any rate, he could never marry any other than a lame woman.

This is followed by the case of a man referred to as "Z.," who,

> even in early childhood always felt great sympathy with the lame and the halt [and] used to limp about the room on two brooms in lieu of crutches, or when unobserved, go limping about the streets . . . in his erotic dreams, the idea of the limping girl was always the controlling element. The personality of the halting girl was a matter of indifference, his interest being solely centered in the limping foot. He never had coitus with a girl thus afflicted. His perverse fancies revolved around masturbation against the foot of a halting female. At times he anchored his hope on the thought that he might succeed in winning and marrying a chaste lame girl. . . . His present existence was one of untold misery.

Finally, there is the case of a thirty-year-old civil servant:

> [S]ince his 7th year he had for a playmate a lame girl of the same age. At the age of 12, puberty set in, and it lies beyond doubt that the first sexual emotions towards the other sex were coincident with the sight of the lame girl. For ever after only halting women excited him sexually. His fetish was a pretty lady who, like the companion of his childhood, limped with the left foot. He sought early relations with the opposite sex but was absolutely impotent with women who were not lame. Virility and gratification were most strongly elicited if the woman limped with the left foot, but he was also successful if the lameness was in the right foot. His sexual anomaly rendered him very unhappy and he was often near committing suicide.[3]

An early cinematic depiction of voluntary amputation appeared in Tod Browning's silent horror film *The Unknown* (1927), starring Lon Chaney as a circus freak named Alonzo the Armless, whose act consists of throwing knives and firing a rifle at his beautiful assistant, Nanon (played by Joan Crawford) with his feet. Alonzo is in fact a fugitive and an impostor, who ties his arms back tightly to conceal a deformed thumb that identifies him. Alonzo secretly loves Nanon, but having been harassed and groped by men for much of her life, she has a morbid fear of their arms. She shuns the advances of Malabar, the circus strongman but, believing Alonzo to be armless, feels comfortable around him. When the circus owner and Nanon's father discover Alonzo's secret, he murders them both. Nanon witnesses the crime, and sees the killer's deformed thumb, but not his face. Alonzo then blackmails a surgeon into amputating both of his arms, so that Nanon cannot identify him as the murderer, and he can win her heart.*

From 1924 to 1941, *London Life* magazine published a long series of letters and short stories, most of them signed by one Walter Stort, containing accounts from young female amputees and men who were sexually attracted to, or romantically involved with, them. In 1972, *Penthouse* magazine began to publish similar letters and stories, which became so popular that the editors started a regular column, "Monopedia Mania," and 1975 saw the publication of an underground pornographic comic, *Amputee Love*, which describes the exploits of Lyn, who discovers a secret world of amputee fetishists and orgies after losing one of her legs in a car accident.[4]

The first case reports of "self-demand amputation" by would-be amputees strengthened this association with sex.** They were published in 1977 in the *Journal of Sex Research* by the renowned sexologist John Money, with Russell Jobaris and the psychologist Gregg Furth, the latter himself a would-be amputee. Money, Jobaris, and Furth considered the condition to be a sexual perversion. Would-be amputees, they wrote, fantasize about having a limb amputated, and fetishize the stump for its resemblance to a phallus. Furth therefore suggested naming the condition "apotemnophilia," meaning "amputation love," to reflect this.[5]

To date, fewer than 500 cases of body integrity identity disorder, as the condition is now called, have been documented in the medical literature. Many of the patients do indeed give sexual reasons for wanting a limb amputated, stating that they find the idea of being an amputee sexually arousing, or that they are sexually attracted to amputees. This is not universal, however, and a large proportion of the documented cases lack a sexual component altogether. In 2005, psychiatrist Michael First of Columbia University published the details of telephone interviews with fifty-two respondents who had BIID. Most stated that their perception of the affected limb was no different from that of their other, opposite limbs; nineteen said that "it felt different in some way"; seven stated that it felt "like it was not

* Spoiler: Afterward, Alonzo, who is now actually armless, returns to Nanon, who tells him that she has finally accepted Malabar's proposal. Brokenhearted, Alonzo tries to maim Malabar during a circus performance, but in the process is trampled to death by a horse.

** More dismissively referred to as "wannabe amputees" or "wannabes" in the literature.

their own"; several reported that that they experienced sensations from the affected limb less intensely; and several more said that the sensations were more intense. Importantly, none of the respondents reported a history of mania, delusions, or hallucinations, and less than one-fifth of them reported sexual arousal as the primary cause of their desire for amputation.

When asked to give the main reason for wanting an amputation, nearly two-thirds of First's respondents stated that it was "to restore" their "true identity." For example, a teacher named Tom told First that he had felt "incomplete" with two arms and two legs. He underwent psychotherapy and took the antidepressants and antipsychotics he was prescribed, but rather than dampen his desire for an amputation, all his treatments just made him feel worse. In 2001, Tom had an elective amputation of his left leg, and this, he reported, finally made him feel "whole" or "complete."

Others with BIID give similar reasons for wanting an amputation. In Melody Gilbert's 2003 documentary *Whole*, an older Florida resident named George Boyer explains that he, too, had "a lifelong obsession" with amputating his leg, and that he had been in psychotherapy his whole life, which had in no way affected his desire to amputate. When he eventually blew his own leg off with a shotgun, he felt "absolutely transformed," adding that he "finally became a person late in life," and that "if I die in the next instant, I don't care, because I have finally realized myself. I have become whole." Another participant in the documentary told the filmmakers that "the paradox is that by taking the leg away, I'm actually more of a person than I was before. . . . In your way of thinking, I've mutilated myself. In my way of thinking, I've corrected the body that was wrong."[6]

On the basis of such reports, First concluded that the condition "may be conceptualized as an unusual dysfunction in the development of one's fundamental sense of anatomical (body) identity," that those with this condition desire amputation in order to "correct a mismatch between the person's anatomy and his or her 'true' self," and suggested that the condition be renamed "body integrity identity disorder."[7]

Although BIID is still largely unrecognized within the medical community, clinical neurologists are familiar with a number of syndromes that present with similar and related symptoms. Patients with somatoparaphrenia deny that a limb belongs to them, and this denial of ownership is often accompanied by a delusional false belief about the limb. They may, for example, insist that the limb belongs to a doctor or nurse in their charge, to their spouse, or, more rarely, that it has been possessed by a dead relative or by the devil. Some patients will claim that the limb has been lost or stolen, that it is horribly disfigured or rotten, or that it has been amputated and replaced with a new limb. Others may perceive their limb as an animal or part of one—a snake or reptile, for example, or a cow's hoof—as an inanimate object, or as an item of food, such as a sausage. Sometimes, these delusions are directly related to the patient's life experience, as in the case of a former slaughterhouse worker who believed that his arm had turned into a slab of dead meat and was hanging at his bedside.[8]

The term "somatoparaphrenia," meaning "body beyond mind," was introduced in the 1940s by the neurologist Joseph Gerstmann to distinguish it from various other syndromes. Gerstmann treated many patients who had had strokes, and he frequently encountered some with strange disturbances of bodily awareness. One common symptom of stroke is hemiplegia, or paralysis on one side of the body, opposite the side of the brain on which the stroke damage has occurred. In some cases, patients lose awareness of the paralyzed limb or limbs, and act as if one-half of their body does not exist. In other cases, they are simply unaware of their paralysis, as Gerstmann explains:

> The patient behaves as though he knew nothing about his hemiplegia, as though it had not existed, as though his paralyzed limbs were normal, and he insists that he can move them and can walk as well as before. Asked to lift up both arms, he naturally moves the healthy one only, but maintains that he has raised the disabled one also. Requests for movements with the paralyzed left arm or leg are performed by him merely with the healthy one, or not at all, but at the same time he is convinced that he has carried out the task. The patient may pay no attention to the paralyzed side, as though he had forgotten it; some not only neglect the defective side, but even refuse to look at it or turn away to the right.[9]

Gerstmann regarded this as "nonperception of disease," and considered the phantom limb to be "the best-known example . . . of nonrealization of one's own defect . . . consisting in the experience of possession of a lost member of the body." But as for stroke patients, unawareness of paralysis may also be accompanied by "various illusions, distortions, confabulations and hallucinatory or delusional ideas," and it is this condition that he named "somatoparaphrenia." Collectively, these syndromes are referred to by the general term "asomatognosia," meaning "loss of knowledge of the body."

When we know something very well, we say that we "know it like the back of our hand." But just how well do you know the back of your hand? You would probably think that you know it very well indeed, and that you couldn't be fooled into believing that a fake rubber hand is yours. Yet a simple and striking illusion suggests otherwise. In the "rubber hand illusion," a test subject sits at a table and places their arms on the tabletop, with their left arm hidden from view behind a screen and with a realistic, life-sized rubber hand placed on the tabletop in front of them (figure 3.1). They sit with their gaze fixed on the rubber hand, while a researcher uses two small paintbrushes to stroke the rubber hand and their real, hidden hand, making sure that the brush strokes are delivered in exactly same way, and at exactly the same time, to each hand.

After a few minutes of this synchronous stroking, the subject is asked to fill out a questionnaire containing open-ended questions designed to determine what they experienced. Most of those tested—about two-thirds—report that the touch sensations they felt came from the rubber hand rather than their real, hidden hand, with eight out of ten spontaneously telling the researchers, "I found myself looking at the dummy hand thinking it was actually my own," or something similar. And, when subjects are asked to close their eyes and use their right hand to point to their left index finger, they tend to point toward the index finger of the fake hand rather than their own.[10]

Crucially, the illusion only works when the rubber hand and the subjects' real hand are stroked in exactly the same way and at exactly the same time. If the brush strokes on their real hand and rubber hand are out of synch with each other, the subjects do not experience the illusion. Likewise, if the rubber hand is rotated by 180 degrees, so that it is in an unrealistic position, the illusion does not occur.

The rubber hand illusion was first described in the journal *Nature* in 1998 and has since been performed on thousands of people in hundreds of follow-up studies, and although the 1998 study used subjective measures to determine subjects' experience of the illusion, these subsequent studies have examined it more objectively. Thus, as a 2006 study showed, if the researcher "attacks" the rubber hand during the illusion, for example, the subject's heart rate suddenly increases, and they start to sweat—involuntary reactions that are characteristic of a fear response. Furthermore, strange things happen to the subjects' real hand while they are experiencing the illusion. Blood flow to the hand decreases, causing its temperature to drop by fractions of a degree, and histamine reactivity—a key step in the immune response—increases. It is as if the brain neglects, or disowns, the real hand, and treats it as a foreign object.[11] Another study in 2019 found that inducing ownership of a rubber hand also appears to make the subjects' real hand less sensitive to pain: subjects take longer to say that an ice pack placed on the back of their hand feels uncomfortable, and the stronger their experience of the illusion, the longer it takes for the cold sensation to feel unpleasant.[12]

There are a number of interesting variations of the original illusion. The same method can be used to induce ownership of a supernumerary third arm or, in the case of the "invisible hand illusion," "a discrete volume of empty space."[13] Several studies show that macaque monkeys can also experience such an illusion, and researchers in Japan have demonstrated that mice are also susceptible to it. To induce the "rubber tail illusion," they

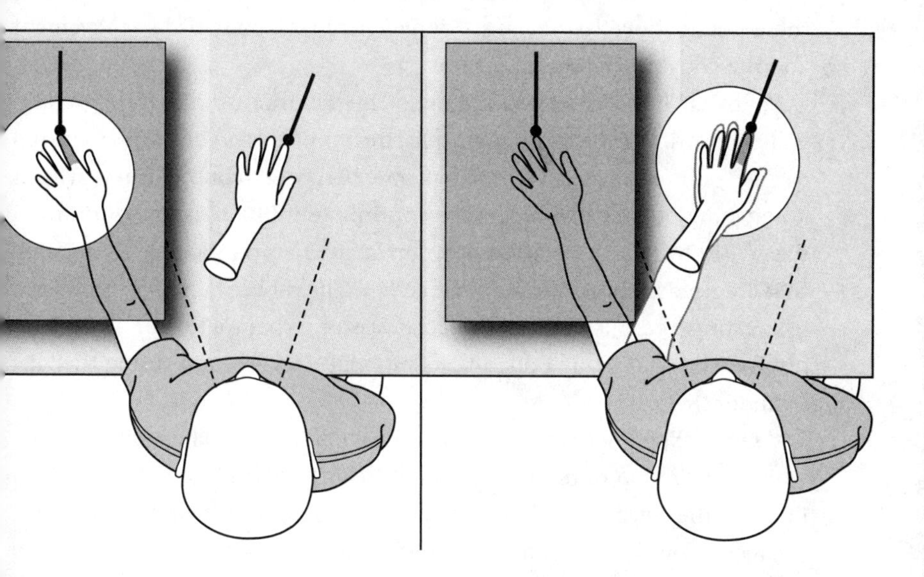

Figure 3.1
The rubber hand illusion. A participant sits at a table, with their left hand hidden from view and the fake hand in front of them in a realistic position. Synchronous stroking of their left hand and the rubber hand gives them the impression that the sensations they feel originate in the rubber hand. Adapted from Metzinger, M. (2008), Empirical perspectives from the self-model of subjectivity: A brief summary with examples, *Progress in Brain Research*, 168:226.

placed mice in a small tube with windows at the front and an opening for the real tail at the back, and placed the fake tail in sight alongside it. When the researchers stroked the real and fake tails synchronously, then grasped the fake tail, the animals reacted as if their real tails were being touched, by turning or retracting their heads.[14]

The rubber hand illusion opens a window onto bodily awareness. It enables researchers to manipulate the sense of body ownership under controlled conditions, and brain scanning studies performed on subjects during the illusion are revealing the underlying brain mechanisms. These studies show that ownership of the rubber hand is associated with increased activity in a brain region called the premotor cortex, with the extent of the increase being closely related to the strength of the illusion. The illusion also seems to decrease signaling between these frontal motor areas and regions of the parietal lobe.[15] Conversely, "disownership" of the real hand during the illusion is associated with reduced activity in the primary motor cortex, which lies immediately behind the premotor cortex.[16] Together, these results show that the sense of ownership is closely tied to movement and the body's potential for action.

All of the brain areas responding in these illusion studies form a brain-wide network that processes multiple streams of sensory information from the body and integrates them to generate bodily self-consciousness. This is an ongoing process that relies on a continuous influx of information from the senses, which the brain uses to update maps and models of the body that are needed for perception and action. The rubber hand illusion works by creating a conflict between what we see and what we feel, and the brain resolves the conflict by updating its internal models, substituting the fake hand for the real one.

Animal experiments reveal something about the cellular mechanisms by which integration of multiple sensory streams occurs. Researchers in 2013 induced the illusion in two macaques by restraining their arms beneath a screen showing computer-generated images of their arms, so that they could see one of the avatar arms being touched by a virtual ball. They also tracked the animals' eye movements and monitored the activity of dozens of individual brain cells with microelectrodes implanted into the motor cortex and the adjacent somatosensory cortex. Both the motor cortex, which controls movement, and the somatosensory cortex, which responds to touch, contain highly organized body maps that guide our actions and localize our sensations. Normally, touch on any given body part will activate the corresponding region of the somatosensory cortex, and a smaller area within the motor cortex. But after a short period of synchronous touch to the monkeys' real and virtual arms, cells in both areas began to respond to virtual touch alone, suggesting that the computer-generated arm had been incorporated into the monkeys' body maps and models.[17]

Clinical examinations show that damage anywhere within this same network of brain structures can cause asomatognosias and various other disturbances of bodily awareness. For example, stroke damage to the right motor and premotor areas of the brain can cause loss of awareness of the left arm and deficits in mental imagery of the limb.[18] These same symptoms can occur as a result of damage to other components of the network, however, and are most often associated with damage to the right parietal lobe, certain parts of which seem to play a special role in bodily awareness. Gerstmann noted that somatoparaphrenia most often affects the left side of the body and is usually associated with damage to "the lower part of the [right] parietal lobe," a region, he added, that "represents the main junction in the functional chain of the process" causing the condition.[19] Contemporaries of Gerstmann, the neurologists Henry Head and Paul Schilder played major early roles in our understanding of bodily awareness; they linked disturbances in how we perceive our body to right parietal lobe damage.

Whereas patients with asomatognosia conditions usually recover spontaneously with time, for those with extreme manifestations of BIID, the desire to amputate a limb does not abate. Neither antidepressants, nor antipsychotics, nor various types of psychotherapy diminish their unremitting desire, and amputation seems to be the only available option to remove that desire, as long as it is done according to their specifications. These would-be amputees know the exact point at which they want their arm or leg amputated, and those who go through with the operation almost always express great relief afterward—unless the amputation is performed even half an inch above or below the location they have specified.

BIID therefore poses something of an ethical dilemma for physicians. In early 2000, it emerged that a consultant surgeon named Robert Smith had amputated the healthy legs of two men in operations carried out in September 1997 and April 1999 at the Falkirk and District Royal Infirmary, a National Health Service hospital in Scotland. "The chairman and board members of Fort Valley Acute Hospitals NHS Trust, which runs the hospital, were unaware of the operations at the time," reported the legal correspondent of the *British Medical Journal*, and "only learnt of them last summer when Mr. Smith informed the trust's new chief executive . . . that he was involved in assessing a third patient." Subsequently, the report continues, "the trust announced a ban on further amputations after a report from its ethics committee," and although "such operations were not ruled out for the future . . . a strict procedure would have to be followed."[20]

On the face of it, amputating a healthy limb does seem to go against the Hippocratic oath to "do no harm." Whereas those having a milder form of BIID seem content just pretending to be amputees by tying back their arm or leg, and using wheelchairs or crutches, those having the most extreme form are determined to be rid of the unwanted limb, and they will go to extraordinary lengths to do so. "Patients . . . often resorted to self-harm," states the *British Medical Journal* report, "for example, by shooting their leg off or lying on a railway track." Fourteen of the fifty-two BIID respondents interviewed for Michael First's 2005 study presented with this "most severe manifestation" of the condition, "which drives [them] to have a surgical or self-inflicted amputation." Of these fourteen, three had convinced a surgeon to amputate, and nine had self-amputated, "using methods that put [them] at risk of serious injury," including using a shotgun, chainsaw, wood chipper, and dry ice," with five of the nine having "amputated one or more [of their] fingers or toes using methods such as a saw, pruning shear, and hammer and chisel." In one well-documented case, a seventy-nine-year-old man from New York traveled to Mexico and paid a back-alley doctor $10,000 to amputate his leg, but died of gangrene a week later.[21]

Paradoxically, then, giving those with BIID a clean surgical amputation would minimize the harm they might cause themselves. Opponents say that those with BIID desire amputation so that they can seek disability benefits, and some members of the disabled community are outraged by the fact that apparently healthy individuals would seek to become disabled. But those who have undergone amputation regard themselves not as disabled, but as "transabled," and none has been known to seek state resources after having a limb removed. On the contrary, elective amputation for those with BIID could be of benefit to society, for their unwanted limbs could be donated to those who need limb transplants.[22] Elective amputation also seems to be the most ethical option with regard to patient autonomy, or individuals' right to make their own decisions about their medical care.

Body integrity identity disorder is currently being considered for inclusion in the *Diagnostic and Statistical Manual of Mental Disorders* (*DSM*). But inclusion in "psychiatry's bible" would have both an upside and a downside. On the upside, it would legitimize BIID as a diagnosable condition, and its entry in subsequent editions of the *DSM* would most likely also include a list of effective treatments. On the downside, inclusion in the *DSM* would classify BIID as a psychiatric condition and thus stigmatize it further. Growing evidence suggests that BIID has a neurological origin, however, and arises because of a long-term disturbance of bodily awareness, in which those having the condition lack a sense of ownership over the affected limb. They do not desire to amputate a healthy arm or leg because they are psychologically disturbed, but rather the reverse: they are psychologically disturbed because they desire to amputate the limb. This desire almost always becomes apparent by early adolescence, and it is likely to arise as a result of events that occurred in early childhood, when bodily awareness is still developing.

4 Representations

Bodily awareness depends upon multiple streams of sensory information from the body into the brain. This information is organized into maps, or "representations," of the body, distributed throughout the brain. In the early part of the twentieth century, the neurologist Henry Head and the psychiatrist Paul Schilder conceptualized the "body schema" and "body image," respectively, based on their observations of patients with brain damage, and the neurosurgeon Wilder Penfield identified systematic representations of the body in the brain. These discoveries and ideas have guided research into bodily awareness ever since. Most researchers agree that the brain contains multiple representations that are critical for how we use, perceive, and think about our body. There is, however, much confusion over the different terms used to describe body representations, and researchers still know very little about how the multiple representations interact to guide our behaviors.

When the Civil War broke out in America, the field of behavioral neurology was gaining ground in Europe. Neurologists observed changes in behavior caused by stroke and other insults and injuries to the brain and then examined their patients' brains post mortem, enabling them to forge links between the brain and behavior. Among the best known of the nineteenth-century neurologists are Pierre Paul Broca and Carl Wernicke, both of whom identified brain areas involved in speech and language. In 1861, Broca visited the Bicêtre Hospital in the southern suburbs of Paris to see Louis Victor Leborgne, a patient who had epilepsy, loss of speech, and right-sided paralysis and who had, by then, been institutionalized for over thirty years. Known throughout the hospital by the nickname "Tan" because this, together with a few obscenities, was the only thing he had said for the past two decades, Leborgne died just six days after Broca's arrival, and when Broca performed an autopsy, he found an egg-sized fluid-filled cavity toward the back of the left frontal lobe of Leborgne's brain. Later that year, Broca saw an eighty-four-year-old laborer with speech disturbances, and subsequently found that he, too, had somewhat more limited damage to the same region of the left frontal lobe. Today, that region is still known as "Broca's area," and the speech disturbances that occur as a result of damage to it are often referred to as "Broca's aphasia." In Germany, at around the same time, Carl Wernicke examined patients who had had strokes but who retained their ability to speak although they could no longer understand the spoken language of others due to damage to another brain region, "Wernicke's area," which lies farther back in the brain than Broca's area, at the top of the left temporal lobe.[1]

Wernicke and others also examined stroke patients who experienced disturbances of bodily awareness. Some lost feeling on one side of their body, for example, or could not recognize objects by touch. Some failed to recognize that one side of their body had become completely paralyzed and clung to the false belief that there was nothing wrong with them. Others lost awareness of one side of their body altogether—they would wash and dress only the unaffected side, shave only on the unaffected side of their face, or even eat from only one-half of their plate.[2]

Such observations led Wernicke to suggest that there is a strong correspondence between the brain and the body. He argued that different body parts send different types of signals to the brain, and that bodily sensations are associated with "memory images." The cerebral cortex integrates sensations and their associated memory images to generate the "somatopsyche" (body consciousness), which he distinguished from the "allopsyche" (consciousness of the outer world) and "autopsyche" (consciousness of the inner self).[3]

Head and Schilder

Bodily awareness research flourished during the early part of the twentieth century. At the London Hospital in Whitechapel, in London's East End, the neurologist Henry Head treated thousands of patients who had sustained damage to their spinal cord or cerebral cortex, many of whom presented with disturbances in how they perceived and experienced their body as a result of that damage. Head used relatively simple methods to examine these disturbances. He gently stroked different parts of each of his patients' body with long wisps of fine cotton wool to test their sensitivity to light touch, and he used carpenter's calipers to determine their ability to discriminate between two simultaneous points of contact. He used heavy tuning forks to test their sensitivity to vibrations, and he rubbed a device called a "Graham-Brown esthesiometer" across their skin to determine their appreciation of roughness. He poked and prodded blindfolded patients, then asked them to indicate on a diagram where they had been touched, and he asked them how their limbs were positioned after they had moved them voluntarily or he had changed their positions himself. He placed test tubes filled with hot and cold water onto various parts of their body, tickled and scraped the soles of their feet, and pricked them with sharp steel pins in order to determine their sensitivity to these different types of sensations.

Head viewed diseases as "nature's experiment[s] on man" and his patients as opportunities to see some of the results of these natural experiments. Some patients were completely insensitive to pin pricks or to light touch, pressure, or hot or cold temperatures on one side of their body. Some could discriminate touch sensations applied at the same time to both an arm and a leg on the affected side of their body but were not aware of the position of their arm or leg, whereas others could not make a two-point discrimination but were aware of the position of their limb on the affected side.

Head took extensive notes during his examinations and compiled detailed case reports of eighteen patients in a 250-page paper, which he wrote with his colleague Gordon Holmes and published in the journal *Brain* in 1911. One of these patients was a thirty-four-year-old man, who "was in perfect health until he was thrown from his van in consequence of a collision on May 30, 1908." First examining the man nearly two years later, Head learned that he had remained unconscious for three weeks after the accident, after which, "he knew he could not recognize the position of the right arm, but was unconscious of the loss of painful and thermal sensibility on the left half of the body." During his examination, Head noted that "all sensibility to touch, pain and temperature was perfectly preserved" on the right side of the body, "but the patient was unable to recognize the posture of his right arm and leg, and could neither name nor imitate correctly the position into which they were placed."

On the basis of observations in dozens of similar patients, Head proposed the concept of the "body schema," which he defined as "a postural model of ourselves." The body schema, he went on to say, "constantly changes," because it records "every new posture or movement . . . and the activity of the cortex brings every fresh group of sensations evoked by altered posture into relation with it," so that "immediate postural recognition follows as soon as the relation is complete." Some patients retained their appreciation of posture while losing the ability to localize touch, however, and Head took this as evidence for a second type of body schema: "A patient may be able to name correctly, and indicate on a diagram or on another person's hand, the exact position of the spot touched or pricked, and yet be ignorant of the position in space of the limb upon which it lies. . . . This faculty of localization is evidently associated with the existence of another schema or model of the surface of our bodies which also can be destroyed by a cortical lesion. The patient then complains that he has no idea where he has been touched."[4]

Head divided his professional time between the London Hospital and his private practice in London's West End, every day traveling back and forth between the two on "the Tube" (London's underground train). In private practice, he worked as a consultant physician. He occupied plush offices on Montagu Square and Harley Street, where he occasionally treated wealthy patients who had multiple sclerosis but, more often, he treated young women who had been diagnosed as "neurotic" or "hysterical," or lawn tennis players experiencing neuritis. Head established a reputation as a specialist in nervous disorders, whose expertise was sought by the rich and famous from far and wide. Among his patients were Virginia Woolf, who consulted him after having a nervous breakdown in 1913; the wife of the English painter and critic Roger Fry when she developed a mental health condition; and the French painter Jacques Raverat, who had multiple sclerosis but stopped visiting Head after several consultations because "he wd. only tell me to rest and I am anyhow spending 18 hours out of 24 in bed."[5]

The London Hospital, by contrast, was located in one of the poorest parts of London. At the time, Whitechapel was populated largely by Eastern European Jewish immigrants, who made up the vast majority of the patients. Head made little attempt to conceal his disgust and hostility toward them—he often complained about the stench of the outpatients' waiting room, and he believed that these patients were unable to properly articulate their symptoms due to their lack of education. Many "are found to be untrustworthy in consequence of misuse of alcohol or other causes," he wrote, and even those he deemed trustworthy can, "at best," answer "only 'Yes' and 'No' with certainty." Such patients, he continued, "can tell little or nothing of the nature of their sensations," so it was "unwise to demand any but the simplest introspection from patients [and] it soon became obvious that many observed facts would remain inexplicable."

Head's solution was to experiment on himself. On April 25, 1903, he traveled with his colleague and collaborator William H. R. Rivers to the home of a surgeon, referred to only as "Mr. Dean," who performed an operation to sever the radial nerve in Head's left arm. Then, once a week for the next four-and-a-half years, Rivers tested the sensitivity of Head's limb to various stimuli. During each session, he outlined the affected areas with a black pen and took photographs to document the gradual return of sensation. In 1908, the pair published their findings in a 127-page paper in the journal *Brain*.

The morning after the operation, "half of the back of [Head's] hand and the dorsal surface of the thumb were found to be insensitive to stimulation with cotton wool, to pricking with a pin, and to all degrees of heat and cold," they wrote in their 1908 paper. "The most striking fact, however, was the maintenance of deep sensibility over the whole of the affected parts on the back of the hand. Pressure with the finger, with a pencil, or any blunt object was immediately appreciated." On June 7, forty-three days after the operation, "the first noticeable change in the extent of the loss of sensation was discovered. . . . The borders of the area insensitive to cotton wool remained unaltered, but the cutaneous analgesia was distinctly less extensive." By September, "the whole forearm had become sensitive to cold [but] still remained insensitive to heat," and by November, "one heat-spot was found near the base of the first phalanx of the thumb."

Overall, "the operation . . . did not interfere with sensibility to the tactile and painful aspects of pressure. But the whole affected area became insensitive to prick, to heat, and to cold. . . . Fifty-six days after the operation (June 20), the analgesia on the forearm had greatly diminished. . . . Eighty-six days after the operation (July 20), the whole forearm responded to prick," and, "Two hundred and twenty-five days after the operation (December 6), the hairs on the back of the hand responded with a diffuse tingling to cotton wool." By November 12, 1904, 567 days after the operation, "the greater part of the affected area on the back of the hand had become sensitive to cutaneous tactile stimuli, and temperatures below 37°C [98.6°F], evoked sensations of warmth."

Head also told Rivers to test the sensitivity of his penis by pricking the glans with needles and dipping it into drinking glasses containing hot and cold water. Their conclusion from that test: the penis "is not sensitive to cutaneous tactile stimuli, but pressure is correctly appreciated and localized with fair accuracy. Sensations of pain . . . are diffuse and more unpleasant than over normal parts. . . . In every case the reaction appears to be more vivid, and yet the glans is entirely insensitive to temperatures between 26°C [78.8°F] and 37°C [98.6°F]."[6]

Meanwhile, at the University of Vienna's Neuropsychiatric Hospital, Paul Ferdinand Schilder examined many patients with similar disturbances in bodily awareness. Schilder built on Head's concept of the body schema and combined it with Wernicke's concept of the somatopsyche to develop the idea of the "body image," defined as "the picture of our own body which we form in our mind [or] the way in which the body appears to ourselves." He described his observations and ideas in a small book, *Das Körperschema* (The schema of the body), published in 1923, which he expanded twelve years later into *The Image and Appearance of the Human Body*, published in 1935. For Schilder, the study of bodily awareness "must be based not only on physiology and neuropathology, but also on psychology"; he conceived of the body image as a multidimensional construct that not only incorporated bodily sensations and how we think about our body, but also took social factors into account.[7]

Schilder left Germany in 1929 for New York, where he taught at New York University and was appointed clinical director of the psychiatric division at Bellevue Hospital. In 1936, he was remarried to child psychiatrist Lauretta Bender, who played an important, albeit largely overlooked, role in early autism research. Four years later, in 1940, a few days after Bender had given birth to their third child, Schilder was knocked down and killed by a car while making his way to visit them in the hospital.[8]

Head is today considered a pioneer of neurology. Schilder, however, remains a relatively obscure figure in the history of the field, even though he is arguably the more influential of the two. Indeed, before his tragic early death, Schilder would author and publish more than a dozen books and some 300 research papers—up to twenty-four per year between 1909 and 1939, except during the First World War—on a wide variety of subjects in neurology and psychiatry.

Schilder considered *The Image and Appearance of the Human Body* (1935) to be his most important work, largely because it helped make the concept of the body image central first to neurology and then to psychiatry. Thus the British neurologist Macdonald Critchley, who went on to become president of the World Federation of Neurology from 1966 to 1973, described Schilder's book as "just . . . short of being one of the great monographs of neurology" in a review of body image research for *The Lancet* in 1950. Noting how "certain regions of the body-image assume particular importance under varying physiological circumstances," Critchley then explained a wide range of conditions and behaviors in terms of modification of the body image: "Any type of pain is liable to cause the affected part to loom large," and in "almost any case of partial paralysis of a limb, the affected segment usually [gives] the impression of being too heavy or too big." The feeling of hunger causes the area around the stomach "to assume importance"; thirst "focuses attention on the tongue and the palate"; and "the sensation of a full bladder causes an enhancement somewhere near the root of the penis in the male, and at a deeper region in the female." He mentioned that drinking too much alcohol "produces complex . . . sensations, ordinarily spoken of as a 'fat head'"; he described "the cult of nudism" as "an interesting psychopathological anomaly" of the body image, associated with heightened perception of the skin. Critchley even speculated on the body image in a two-headed "human monster," suggesting that it would be, "to say the least, complicated."[9]

By the 1960s, the body image had become central to psychology and psychiatry, in large part due to the work of Seymour Fisher and Sidney Cleveland. Believing that how people perceive and experience their body is closely related to, and can even predict, their personality and behavior, Fisher and Cleveland developed the "body image boundary" as an index of how people perceive their body in relation to the outside world. An individual's "Barrier score" was determined by their responses to inkblot tests, serving as an index of the extent to which they perceive their body as a protective covering. Thus tabulating responses such as "turtle with a shell," "cave with rocky walls," or "entrance to a house" gave a measure of how protective or easily invaded an individual's body boundary was. In a series of studies, reported in 1956 and 1964, Fisher and Cleveland found major differences in how people perceive their body boundaries. People with a definite boundary emphasized its protective nature and are more likely to "behave autonomously, manifest high achievement motivation, be invested in task completion, be interested in communicating with others, and serve an active integrative role in small groups." People in this "high-barrier" group "are more likely to take a 'muscular' attitude toward life, emphasizing values having to do with getting to the top."[10]

Psychiatrists began using the concepts of the body image and body boundary to explain conditions such as schizophrenia, which came to be viewed by some as a disturbance in which, according to one 1960 report, "the boundaries of the self become loose or blurred," such that "the patient may feel . . . that parts of his body do not belong to him or that he is part of the plants, animals, clouds, other people or of the whole world and that they are part of him." Case studies of the time describe a wide variety of body image disturbances in patients diagnosed with schizophrenia, including distortions in perception of the size and shape of their body parts, arising due to "certain misinterpreted bodily sensations combined with hallucinatory experiences of some special senses, e.g., of smell." Among these are several cases of "zoophilic metamorphoses," such as those of a thirty-eight-year-old park ranger who reported that his "body had a *doggy smell* [that] everybody could smell," who then "began to have the *feeling* that my body shrank and changed into the body of a dog"; a thirty-nine-year-old Jamaican crane operator, who "began to hear 'voices' alleging that he had 'crab hands'" and "also had the *feeling* in my hands that they shrank and became hard like crab's claws"; and a forty-seven-year-old woman who was "distressed by a '*penetrating smell*' of a cat emanating from her body," who reported that "I also *feel* that my hands and feet shrank, became covered with hair and turned into cat's paws." A 1967 report describing such patients concluded that approximately one-quarter of psychiatric patients experience some kind of body image disturbance, noting that schizophrenia is most often "accompanied by bizarre disturbances in the *shape* of the Body Image," whereas depression is also linked to "disturbances in the *mass* of the Body Image."[11]

A wide variety of neurological and psychiatric conditions—from phantom limb phenomena, somatoparaphrenia, and body integrity identity disorder (BIID) to more common conditions such as anorexia nervosa and body dysmorphic disorder (BDD)—can be thought of as "body image disorders," in which an individual's perception of the body is distorted or otherwise disturbed, such that it does not match the actual physical form of the body. Schilder's influence extends far beyond the mind and brain sciences, however. The body image concept is used widely in the humanities and has also entered into the popular imagination. Indeed, multidisciplinary body image research even has its own peer-reviewed academic journal, *Body Image*, launched in 2004. In an editorial in the first issue, founding editor-in-chief Thomas Cash charts the steady growth of body image research in the latter part of the twentieth century—searches of the PsycINFO and PubMed databases of scientific literature for papers containing the term "body image" yielded 726 results from the 1970s, 1,428 from the 1980s, and 2,477 from the 1990s. Cash invited academics to submit original research papers, theoretical articles, and practitioner reports on a wide-ranging variety of topics: the effects of specific physical characteristics, such as body size, attractiveness, and appearance on body image, psychological functioning, interpersonal processes, and quality of life; cross-cultural and ethnic studies of physical appearance and body image; studies of physical appearance and body image in all medical and health contexts, including cosmetic and reconstructive surgery, dentistry, dermatology, endocrinology, neurology, nursing, obstetrics and gynecology, physical therapy and rehabilitation; the relationship of body image to exercise, eating and dieting, and grooming and other appearance-modifying behaviors; and conceptual contributions to body image research from the evolutionary, feminist, and psychodynamic perspectives, among others.[12]

The Little Man in the Brain

Although the body schema and body image help to explain how the brain represents the body and how representations guide behavior, they are both hypothetical. Our understanding of how the brain actually represents the body comes largely from the pioneering work of the Canadian neurosurgeon Wilder Penfield.

In the second half of the nineteenth century, investigators had started to establish the idea that certain parts of the brain are responsible for particular functions, such as speech and movement. During the 1860s, at what is now the National Hospital for Neurology and Neurosurgery in London's Queen

Square, John Hughlings Jackson examined hundreds of epilepsy patients who were experiencing seizures or paralysis. He became fascinated by how seizures often caused a typical sequence of convulsions that "marched" up or down one side of the patient's body, reporting in 1868 that "I think the mode of beginning makes a great difference as to the march of the fit. When the fit begins in the face, the convulsion involving the arm may *go down* the limb. . . . When the fit begins in the leg, the convulsion marches up; when the leg is affected after the arm, the convulsion marches *down* the leg"; and he noted that, "in very many cases of epilepsy . . . the convulsions are limited to one side of the body; and, as autopsies . . . appear to show, the cause is obvious organic disease on the side of the brain, opposite the side of the body convulsed, frequently on the surface of the hemisphere."

In a landmark 1870 paper titled "A Study of Convulsions," Hughlings Jackson proposed that seizures originate in the cerebral cortex, as a result of "an excessive, and a disorderly discharge of nerve tissue on muscles," and his clinical observations led him to deduce the existence of a motor brain region containing a body map that was organized "somatotopically"—that is, with the spatial relationships between body parts being maintained, such that adjacent body parts are located next to each other on the map.[13]

At around this time, the anatomist Gustav Theodor Fritsch and the psychiatrist Eduard Hitzig provided the first direct evidence for a motor region in the cerebral cortex organized in this way. As a military surgeon in the Franco-Prussian war, Hitzig had treated soldiers with severe head injuries, and he noted that weak electrical stimulation applied to the back of their heads sometimes produced eye movements. Working in Berlin, when he and Fritsch electrically stimulated the exposed brain of a dog, they found that low levels of stimulation applied to the back of the frontal lobe produced isolated movements in individual muscle groups on the opposite side of the body: "A part of the convexity of the hemisphere of the brain is motor, another part is not motor," they wrote in an 1870 paper describing their work. "The motor part, in general, is more in front, the non-motor part more behind. By electrical stimulation of the motor part, one obtains combined muscular contractions of the opposite side of the body." They also found that surgical removal of the left "motor part" elicited contraction or flexion of the right forepaw, resulting in abnormal posture and impaired movements: "In running the animals used the right forepaw wrongly, sometimes more inward and sometimes more outward than the other ones, and frequently glided out with this paw, never with the other ones, so that they fell to the ground . . . the paw is put down with the [wrong] side without the dog noticing it. In sitting on the hind legs . . . the right front paw gradually glides away toward the outside until the dog lies on the right side." The experiment unequivocally showed that nervous tissue is "electrically excitable," or can be electrically stimulated, and by doing so, established the field of neurophysiology. It also provided clear evidence that the frontal lobe contains an area devoted to motor function.

The Scottish neurologist David Ferrier replicated and built on these findings in a series of studies performed in 1876 on different animal species, including dogs, cats, rabbits, and jackals. When he used an induction coil to apply alternating current to the cerebral cortex for longer periods of time than Fritsch and Hitzig, he evoked not only movement in different body parts by stimulating different parts of the cortex, but also finer responses, such as muscle twitches in the eyelid or ear, and movement of a single digit. In this way, Ferrier identified "centers" within the cortex related to fine movements in different parts of the face and body and noted how similar their organization was to the somatotopic map proposed by Hughlings Jackson.[14]

The human brain was still largely uncharted territory, however, and Wilder Penfield would become its master cartographer. He developed a method for electrically stimulating the brain in conscious patients, taking "intriguing expeditions" into this strange terrain, during which he was able to map sensory, motor, and other functions to distinct locations within its peaks and valleys.

In 1934, Penfield established the Montreal Neurological Institute (MNI), where he would operate almost exclusively on patients with epilepsy. At around this time, the anticonvulsant phenytoin was made available, although it proved to be ineffective in a small minority of patients. Because frequent and severe seizures are debilitating, surgery is the only option for those who are resistant to anticonvulsants. But the brain is an extremely delicate and complex organ, with tens of thousands of cells packed into every cubic millimeter of tissue, thus surgery runs the risk of damaging regions involved in functions such as speech and movement. To minimize that risk, Penfield kept his patients conscious throughout their surgeries. After injecting local anesthetic into a patient's scalp, he would open the patient's skull to expose the surface of the brain and, before removing the abnormal tissue causing the seizures, he would electrically stimulate the areas around it, and observe any adverse effects on the patient. He might, for example, ask a patient to count up to ten while he stimulated tissue near the brain's speech centers; if at any point his electrodes interfered with the patient's ability to speak, he knew that this particular patch of tissue was essential for speech, and he steered his scalpel clear of it. Penfield was thus able to locate the specific tissue source ("locus") of the patient's seizures and remove it without inflicting collateral damage. During his operations, he would systematically stimulate different parts of the patient's brain, labeling each small patch of tissue with a numbered sticker and noting the responses evoked by stimulation (figure 4.1). Some regions elicited the recall of long-lost memories; others triggered musical hallucinations; and still others evoked smells.

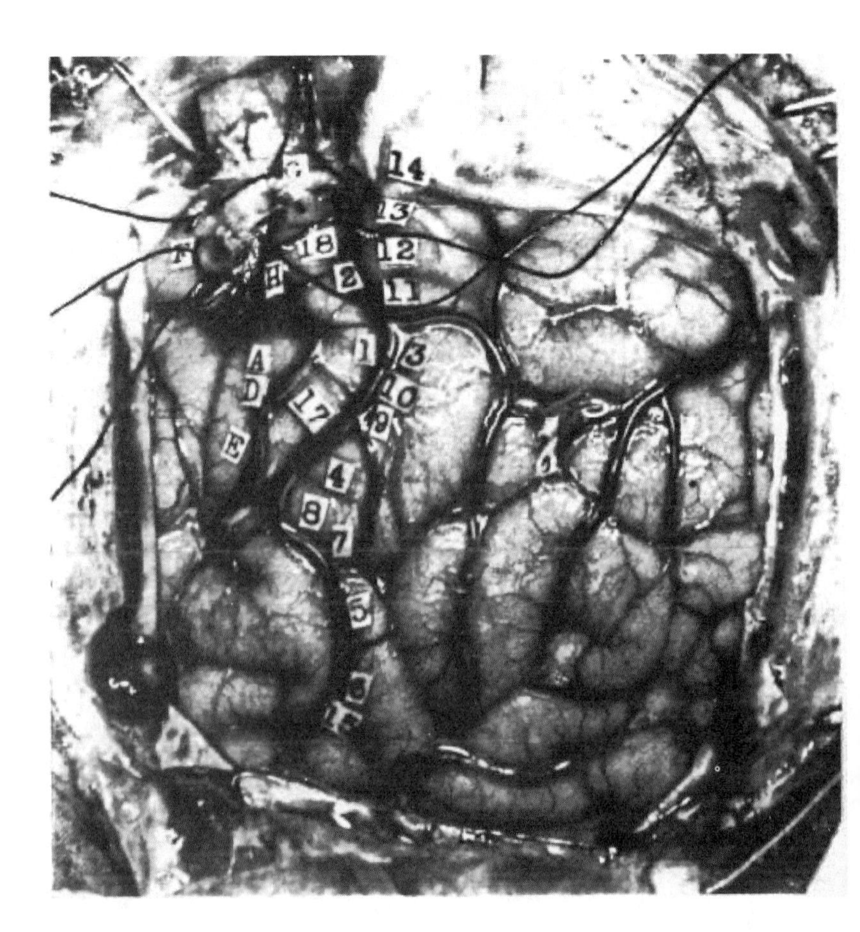

Figure 4.1

Cerebral cortex of "F.W.," a young male whose "focal epileptic seizures [were] characterized by sudden feeling in the right side of the body and movements of the right hand." Penfield used microelectrodes to stimulate the cerebral cortex in conscious patients and placed numbered stickers on the surface of a patient's brain to record the response elicited by stimulation of each part:

14. Tingling from the knee down to the right foot, no numbness.
13. Numbness all down the right leg, did not include the foot.
12. Numbness over the wrist, lower border, right side.
11. Numbness in the right shoulder.
3. Numb feeling in hand and forearm up to just above the forearm.
10. Tingling feeling in the fifth or little finger.
9. Tingling in first three fingers.
4. Felt like a shook and numbness in all four fingers but not in the thumb.
8. Felt sensation of movement in the thumb; no evidence of movement could be seen.
7. Same as 8.
5. Numbness in the right side of the tongue.
6. Tingling feeling in the right side of the tongue, more at the tip.
15. Tingling in the tongue, associated with up and down vibratory movements.
16. Numbness, back of tongue, mid-line.
(G) Flexion of knee.
18. Slight twitching of arm and hand like a shock, and felt as if he wanted to move them.
2. Shrugged shoulders upwards; did not feel like an attack.
(H) Clonic movement of right arm, shoulders, forearm, no movement in trunk.
(A) Extreme flexion of wrist, elbow and hand.
(D) Closure of hand and flexion of his wrist, like an attack.
17. Felt as if he were going to have an attack, flexion of arms and forearms, extension of wrist.
(E) Slight closure of hand; stimulation followed by local flushing of brain; this was repeated with the strength at 24. Flushing was followed by pallor for a few seconds.
(B) Patient states that he could not help closing his right eye but he actually closed both.
(C) Made a little noise; vocalization. This was repeated twice. Patient says he could not help it. It was associated with movement of the upper and lower lips, equal on the two sides

Reproduced from Penfield, W., and Boldrey, E. (1937), Somatic motor and sensory representation in the cerebral cortex of man as studied by electrical stimulation, *Brain*, 40:425–427, with permission from Oxford University Press.

Penfield's method came to be known as "the Montreal Procedure," and is still used regularly in neurosurgical assessment. Stories about patients undergoing "awake brain surgery" make the headlines regularly—in 2015, for example, news outlets worldwide reported that twenty-seven-year-old jazz musician Carlos Aguilera played his saxophone during a twelve-hour operation at the Regional Hospital of Málaga to remove a brain tumor and, in 2020, they reported on Dagmar Turner, a fifty-three-year-old management consultant from the Isle of Wight, who played her violin while neurosurgeons at King's College Hospital in London removed a tumor from her right frontal lobe.

The living human brain looks like a lump of jelly with few recognizable features on its surface. Viewed from above, one important landmark is the central sulcus, a deep fissure running across the middle of the outer surface of each of the brain's two hemispheres, which separates the frontal lobe of the brain from its parietal lobe. The motor cortex is a narrow strip of tissue, running down from top to bottom on the surface of each hemisphere and lying immediately in front of the central sulcus; the somatosensory cortex, a somewhat wider strip of tissue also running down the surface of each hemisphere, lies parallel to and immediately behind the central sulcus (figure 4.2). During operations on hundreds of patients, Penfield found that electrical stimulation of the tissue just in front of the central sulcus on one hemisphere of the brain produced small movements or muscle twitches in specific parts of the opposite side of the body, whereas stimulation of the tissue just behind the central sulcus evoked touch sensations at specific locations. Crucially, although the precise location of areas that evoked specific movements or sensations differed widely between patients, the sequence of responses was always exactly the same. Stimulation down inside the longitudinal fissure separating the two hemispheres of the brain evoked movements or sensations in the genitals, while stimulation progressively farther up toward the midline of the brain elicited responses in the toes, foot, and then leg. Electrical stimulation at the very top of the brain evoked movement or sensation in the hip and torso; moving the electrode progressively farther down along the outer surface of the brain elicited responses first in the shoulder, then in the arm, elbow, forearm, and wrist. This was followed by a large brain area devoted to the hand, with adjacent representations of each of the digits, followed by another large brain area devoted to the face, tongue, and throat, located down near

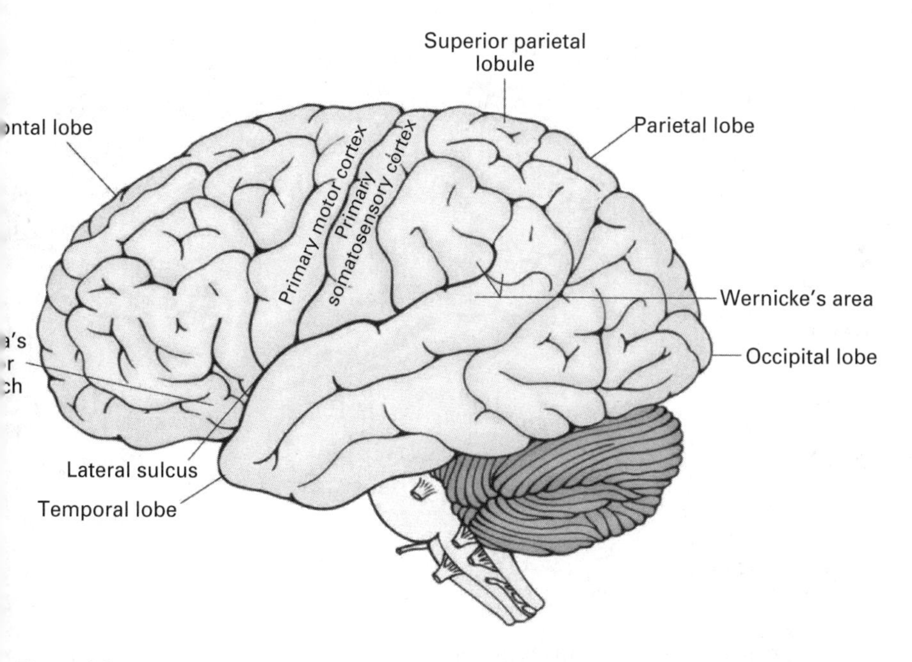

Figure 4.2
Gross anatomy of the brain (viewed from left side)

the lateral sulcus, which separates the frontal and parietal lobes from the top of the temporal lobe.[15]

Penfield visualized these findings in the form of a "homunculus" atop the sensory and motor regions of the brain. This "little man" was born in a pathbreaking 1937 paper Penfield coauthored with his colleague Edwin Boldrey as a naked male figure with outsized hands sitting upside down above his disembodied face, tongue, and throat. Penfield then commissioned medical illustrator Hortense Cantlie to redraw the homunculus, and her diagram, showing the disproportionately sized body parts splayed out over a cross section of the brain, appeared in his 1950 book *The Cerebral Cortex of Man*, coauthored with his colleague Theodore Rasmussen (figure 4.3).[16] Since then, the homunculus has also been represented as a three-dimensional clay model with an enormous head and hands attached to a tiny body, versions of which are on display at the Natural History Museum in London and elsewhere. Penfield explained that "the size of the parts of this grotesque creature were determined . . . by the apparent perpendicular extent of representation of each part." The face and hands are

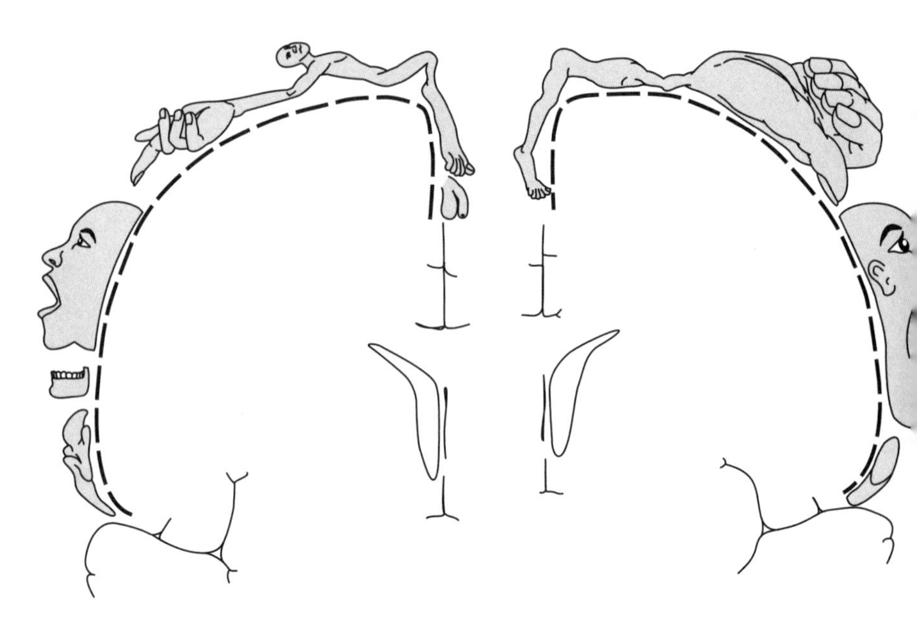

Figure 4.3
Penfield's homunculus. Adapted from Penfield, W., and Rasmussen, T. (1950), *The Cerebral Cortex of Man: A Clinical Investigation of the Localization of Function*, New York: Macmillan, 56.

the most highly articulated parts of the body—the face contains 43 muscles which control our facial expressions and other movements, and each hand contains 29 bones attached by 123 ligaments to 17 muscles in the palm and 18 more in the forearm. This requires a lot of neural real estate—approximately half of the motor cortex is devoted to the face and hands. Similarly, the face and hands are the most sensitive parts of the body, with a far higher density of nerve endings than others, and so approximately half of the sensory cortex is devoted to them, too.

Penfield confirmed and extended the earlier work in humans by showing that the human cerebral cortex contains a region devoted largely to movement and another largely devoted to sensation, and that both of these brain regions contain topographical maps of the body, in which the spatial relationship between representations of body parts in the brain regions corresponds closely to the physical locations of the body parts on the body—a correspondence referred to as "somatotopy." Penfield and Boldrey's use of a diagram to illustrate their findings in 1937 was somewhat revolutionary: until then, investigators had written detailed descriptions of the effects produced by electrical stimulation of the brain; the 1937 homunculus was the first attempt to illustrate cortical representations pictorially, and it set the precedent for experimental brain work to come. Subsequent studies of cortical representations in monkeys were illustrated analogously (as a "simiusculus"), and so, too, were studies carried out in rats (as a "rattunculus").[17]

Somatotopy is a common feature of the mammalian brain, with each species having topographic maps that are organized along the same principle but specialized to reflect the anatomy, lifestyle, and behavior of its members. In rats, for example, a large proportion of the somatosensory cortex is devoted to processing inputs from the whiskers, and this region contains clusters of cells that have a topographical arrangement that mirrors the arrangement of whisker follicles on the rat snout, with each cluster representing an individual whisker; in bats, large parts of the somatosensory cortex are devoted to the head, feet, and wings, but the representations of individual digits within the bat wing area are found side by side, rather than one on top of the other, reflecting modification of the body plan and forelimb for flying; and in the star-nosed mole, nearly half of the somatosensory cortex is devoted to the twenty-two fleshy appendages that surround the mole's nostrils, with the cells organized in a starlike arrangement of stripes, each representing an individual nostril appendage.[18]

Penfield's observations also revealed that the somatosensory and motor brain areas "overlap each other consistently and correspond to each other horizontally," and this has been confirmed in a 2017 study of white matter distribution within the primary motor and somatosensory areas. White matter consists of nerve fibers that transmit information between distinct areas of the brain, called such because the fibers are insulated by a fatty material called "myelin," which appears white under the microscope. Using several different brain scanning techniques, researchers at the Max Planck Institute for Human Cognitive and Brain Sciences in Leipzig found that both brain areas are subdivided into distinct fields, each representing a major body part, including the hand and the face, and are demarcated from one another by myelin boundaries. Furthermore, corresponding fields on either side of the central sulcus appeared to be strongly connected to each other, indicating that they signal each other more than they do other body part subfields within the same vertical strip of tissue. Traditionally, the motor and somatosensory strips were thought of as separate brain areas with distinct functions. But these findings suggest that the corresponding subfields in each—such as the face and hand representations—form a single, functional unit spanning the central sulcus, leading the researchers to conclude that the layout of body parts may play an important role in the organization of brain architecture during development. Their findings support the idea that the somatosensory and motor functions of the brain are closely linked to each other.[19]

Some have pointed out that the homunculus tells us nothing about how the female body is represented in the brain. Although Penfield did speculate about representations of the female anatomy, fewer than a tenth of the more than 1,000 patients he operated on were women, and his case studies include only a few examples of female patients who reported sensations in their breasts and genitalia. So perhaps it's not surprising that the homunculus is obviously male, with representations of the penis and testes—and with no female anatomical parts.

There is, however, reason to believe that the female and male bodies are represented differently. Even though the clitoris is anatomically homologous to the penis, several studies examining the brain's response to clitoral stimulation have mapped the representation of the clitoris to a slightly different location within the somatosensory cortex than that of the penis, and there is some evidence that the clitoris, cervix, and vagina each have distinct representations. Together, findings such as these are beginning to show what a somatotopic map of the female body—displayed by a "*her*munculus"—might look like. Greater knowledge of how the female body is represented in the brain could improve our understanding of how the representations, and the bodily sensations associated with them, change in response to experiences such as hysterectomy, mastectomy, menopause, and pregnancy.[20]

Learning more about how body representations differ between the sexes could contribute to our understanding of the biological basis of gender identity. One study in 2008 surveyed transgender individuals on the incidence of phantom breast and penis sensations following gender reassignment surgery. Of the 29 female-to-male transgender respondents, all had undergone chest reconstruction surgery, but only 9 had started or completed the multistage procedure to reassign their genitals. Even so, 18 of the 29 respondents reported having felt vivid sensations of a phantom penis for many years, whereas just 3 reported experiencing phantom breast sensations postoperatively. Similarly, just 6 out of the 20 male-to-female transgender individuals who took part reported experiencing phantom penis sensations after surgery. A mismatch between the body and its representations in the brain could contribute to gender dysphoria, the feeling of being "a man born into a woman's body," or vice versa. Preliminary evidence for this hypothesis comes from a small 2017 study that used magnetoencephalography (MEG) to record brain responses to tactile stimulation of the breast and hand in 8 female-to-male transgender individuals and 8 nontransgender female controls. The transgender participants exhibited significantly reduced activation of the somatosensory cortex in response to touch sensations on the breast, but not on the hand, compared to the nontransgender controls. Cortical representation of the breast may therefore be diminished or absent, and this could result in a reduced sense of ownership of the breasts.[21]

Remapping

Historically, it was believed that mammals are born with all the brain cells they will ever have, that the adult mammalian brain is a fixed structure, and that damage to the brain is irreparable. In 1913, Santiago Ramón y Cajal, the father of modern neuroscience, wrote that "once development was ended [in adulthood], the founts of growth of the axons and dendrites dried up irrevocably. In the adult centers, the nerve paths are something fixed, ended and immutable. Everything may die, nothing may be regenerated."[22] Consequently, the view that the adult mammalian brain is "immutable" became a central dogma of neuroscience.

Beginning in the 1960s, however, evidence began to emerge that this is not the case. Researchers observed the growth of new cells in the brains of adult rats, cats, dogs, and monkeys (as well in the brains of adult nonmammals such as songbirds)—and, eventually, in the adult human brain, too. Now we know that the brain is in fact highly dynamic. New brain cells are born, and old ones wither away; neurons modify the synaptic connections they form with one another, to make them stronger or weaker. What's more, brain cells can extend and retract their fibers or grow entirely new ones, such that the brain can "rewire" itself to compensate for cortical injuries and insults. Such changes in brain cells, collectively referred to as "neuroplasticity," are the rule rather than the exception. Indeed, plasticity is an inherent property of all nervous systems: the brain evolved to adapt to new circumstances and optimize its performance accordingly. In other words, our brain modifies both its structure and functions continuously in response to every experience we have. These changes begin in the womb and continue throughout our life.

Because of the prevailing view at the time, Penfield did not consider the possibility that the body maps in the somatosensory and motor cortices might change. But plasticity occurs in nerve cells at all levels of organization: in the activity of genes and molecules within nerve cells and the structure of the cells themselves, in local networks of neurons, and in the brain's long-range pathways and overall anatomical structure. The representations of the body are also subject to plasticity—and so, too, is our self-perception.

That body maps change in response to experience started to become apparent in the early 1980s, beginning with a series of animal studies that essentially repeated Henry Head's self-experiments. In 1983, Michael Merzenich of the University of California, San Francisco, and his colleagues severed the median nerve in adult owl monkeys and squirrel monkeys and used microelectrodes to examine the responses of single nerve cells in the somatosensory cortex. This revealed that the monkeys' hand map gradually reorganized itself over a period of two to nine months following the operation, so that nerve cells that had previously responded to touch stimulation on those parts of the hand supplied by the median nerve now responded to touch on adjacent areas supplied by the ulnar and radial nerves instead.[23]

In the experiments of a 1988 follow-up study, Merzenich and colleagues amputated one or two fingers in several adult monkeys and found that the representations of the adjacent digits expanded to occupy the brain areas that the amputated digits previously had. These experiments revealed the apparent limits of this expansion—reorganization occurred up to between .5 and .7 millimeters beyond the boundaries of the deprived cortical area, but no farther.[24] By contrast, in 1990, Merzenich and colleagues found that surgically fusing two fingers together blurred the boundary between the representations of each finger in the hand area of the body map in the monkeys' brains, such that the adjacent regions that had previously responded to touch stimulation applied to one finger or the other now responded to touch on both digits.[25] Finally, in 1999, Jacques-Olivier Coq and Christian Xeri, working with adult rats in Marseille, France, found that decreasing the amount of touch stimulation to a given body part alters the size of the corresponding body map area accordingly. Whereas other researchers found that training monkeys to obtain food pellets by placing their fingertips on a rotating disk with an alternating pattern of raised and lowered surfaces led to expansion of the finger representations in the monkeys' somatosensory cortex, Coq and Xeri found that immobilizing adult rats' forepaws with a cast for a week or two decreased the size of the forepaw representation in the rats' somatosensory cortex by about half.[26]

Brain scanning technology confirms that many of these findings hold true in humans. For example, violinists and other string players use the second to fifth digits of their left hand to finger the strings of their instruments. This requires a great deal of manual dexterity and enhances the touch sensations on these four fingers, which is associated with an expansion of the cortical representations of the fingers. A 1995 brain scanning study of six violinists, two cellists, and one guitarist showed that the extent of expansion is related to the amount of a subject's experience as a musician, with musicians who started playing at an earlier age having larger finger representations than those who started later. Representations of the left thumb, which is used to grasp the neck of the instrument, and of the fingers of the right hand, which manipulates the bow, remained largely unchanged, and were similar to those representations in nonmusicians.[27]*

Temporary immobilization of a limb has the opposite effect. Wearing a cast for a few weeks shrinks cortical representations of the fingers, reducing their activity in response to touch and impairing their touch acuity, compared to size and levels measured in controls, whose corresponding limb was not immobilized and who did not wear a cast. In a 2009 study, such shrinkage, reduction, and impairment were found to be reversible, however—the changes in both brain organization and perception were reversed two to three weeks after the cast was removed, with representations of the fingers returning to their former size and responsiveness to touch as well as touch acuity returning to their former levels.[28] Such changes can occur over far shorter periods of time. In a more recent study, in 2016, researchers found that joining participants' right index and middle fingers together with surgical glue for just one day blurred representations of the

* This refers to right-handed musicians and left-handed musicians who play right-handed, but not to left-handed musicians who play left-handed.

ring finger with those of the little finger, reducing the touch sensitivity of these two fingers.[29]

Congenital amputees—people born missing one or more limbs or other body parts—exhibit more dramatic alterations. A 2017 study of people born missing a hand found that the somatosensory brain area that normally would represent their missing hand represents instead other body parts used to compensate for the missing limb, such as their existent hand or arm, their feet, and even their lips, and a 2019 study of congenital bilateral upper-arm amputees who use their feet to paint found that the foot areas of their brain's body map are enlarged and contain an ordered, topographic representation of the toes that resembles the hand representation in nonamputees.[30]

Reorganization of somatotopic body maps may explain some of the characteristics of phantom pain and other phantom sensations. An influential 1995 study examined remapping in thirteen upper-limb amputees with varying degrees of phantom pain and found a strong correlation between the amount of reorganization and the magnitude of pain reported, with those who reported more extreme phantom pain exhibiting a greater extent of reorganization. Remapping may therefore be a maladaptive form of neuroplasticity, with phantom pain being a consequence of the brain's adaptive response to amputation.[31] Furthermore, in arm amputees, different types of sensations arising in the lower face are sometimes perceived as originating from their missing limb. This may be because the face region of the body map expands and invades the neighboring hand region, which, despite now being deprived of sensory inputs, still retains a map of the hand. Thus the missing hand brain area is said to be "remapped" onto the face, and stimulation of the face can evoke parallel sensations on both the face and the missing hand.[32] In support of this hypothesis, one case study in 1994 showed that the face can also be remapped onto the hand—it described a female patient who had two branches of the fifth cranial nerve removed from the right side of her face and subsequently reported that touch applied to her hand sometimes evoked sensations on her face.[33]

More recent studies challenge the remapping hypothesis, however. One of these, a 2013 study, used functional magnetic resonance imaging (fMRI) to scan the brains of eighteen noncongenital upper-limb amputees, eleven congenital one-handed amputees, and a number of two-handed controls, to show that almost all of the amputees exhibited significant activation of the missing hand area of the brain's body map while moving their phantom hand. Furthermore, a history of greater phantom pain was correlated with greater amounts of activity in the missing hand area, suggesting an increased representation of the missing hand. This study further revealed a reduction in the number and strength of connections between the missing hand area and other areas of the brain's body map in amputees, including the hand area on the opposite side of the brain. This, too, was correlated with phantom pain, with amputees who experienced greater pain exhibiting a greater reduction in connectivity.[34]

A 2015 follow-up study showed that in seventeen arm amputees, the lip area of the brain's body map expands toward the area of the missing hand but does not invade it. Another in 2016 used a high-resolution 7 Tesla scanner to examine the missing hand area of the body map in amputees, revealing that, even decades after amputation, the missing hand representation remained largely intact, though slightly weaker and with greater overlap between the representations of individual digits. As an alternative to the remapping hypothesis, the authors of these studies suggest that phantom pain acts to preserve representation of the missing limb, while also isolating it from the rest of the body map and from other somatosensory and motor representations by disturbing the long-range connections between them.[35] In line with this alternative hypothesis, another team of researchers reported in 2012 that leg amputees exhibit reduced amounts of white matter in the corpus callosum, the massive bundle of nerve fibers connecting the left and right hemispheres of the brain.[36]

The somatotopic organization of the body map has also been used to explain why some people have a foot fetish, and reorganization of the map may explain why foot binding was popular in medieval China. Penfield observed that the somatotopic representations of the feet and the genitals are next to each other, both of them being located near the top of the somatosensory cortex in the region lying inside the long crevice separating the two brain hemispheres. This closeness may result in "cross-wiring" between the two brain areas, leading some people to be sexually aroused by feet. The custom of foot binding in Chinese girls is thought to have started sometime around the tenth century CE. The practice involved tightly wrapping the feet of young girls from about six years of age to prevent the feet from growing to full size, which was believed to enhance the girls' later prospects of marriage. Indeed, according to historians, Chinese men believed that foot binding made a woman's vagina more muscular and sensitive, and they therefore preferred women with bound feet as sexual partners. There may be a scientific basis to this belief: because foot binding may prevent proper development of the somatosensory representation of the feet, causing the brain's representation of the genitalia to expand, women with bound feet might well be "more sensitive and pleasurable lovers."[37]

Injury to the brain can alter its body map topography even more dramatically. In 2009, a team of researchers and clinicians in Switzerland had the rare opportunity to visualize the brain activity associated with a "supernumerary phantom limb." A sixty-four-year-old woman was admitted to the hospital after experiencing both a sudden weakness on the left side of her body and slurred speech, and neurological evaluation revealed that she had had a stroke underneath the cerebral cortex on the right side of the brain. Four days later, the woman reported the appearance of a "pale, milk-white and transparent" phantom third arm protruding from her left elbow. She claimed that she only experienced the phantom arm when she intentionally "triggered" it, whereupon she could see it, move it, and use it to touch certain parts of her body. When she clenched her left hand, she said, she could feel her phantom hand clenching, too; when she touched her body, she felt sensations in the phantom hand as well as on the touched body part; and she felt a sense of relief whenever she used the phantom hand to scratch an itch.

Brain scanning, performed three weeks after the patient's stroke, revealed that the imagined movements of the extra phantom limb activated the hand regions of the right somatosensory and motor cortices, along with other brain areas in the right temporal and parietal lobes known to contain body representations. Furthermore, when the patient used the phantom hand to scratch her right cheek, the scanner also detected activity in the visual cortex and left somatosensory cortex, consistent with the patient's claims that she could see the phantom hand and could feel relief when she used it to scratch an itch.

Stroke interrupts the flow of blood to areas of the brain that receive their supply from the affected blood vessels. This interruption can be temporary or permanent, and the extent of damage caused is directly related to the length of time the blood flow is interrupted. In this unique case, the stroke affected parts of the brain that encode its body map, seemingly causing the arm region of the map to be reduplicated. The woman's supernumerary phantom persisted for four months of rehabilitation, after which it began to diminish, as sensation and movement gradually returned to her left arm.[38]

How Many Body Representations?

The brain contains multiple representations of the body, which are crucial for both action and perception, but it is unclear how different body representations interact with one another, or exactly how they relate to conscious awareness of the body. Some researchers make a distinction between an implicit (or unconscious) body schema and an explicit (or conscious) body image, but there is much confusion over the terms "body schema" and "body image," and this dates back to their earliest usage—Schilder himself often used the two terms interchangeably. Some modern researchers suggest that the terms are no longer useful and should be abandoned altogether, and some simply avoid them, using the more general term "representations" instead.

Other researchers have elaborated on the original concepts. Various studies show that primary somatosensory representations do not always directly correspond to perceived sensations. Different regions of the skin surface have different levels of touch sensitivity, reflecting both the density of nerve endings in any given skin region and the size of its corresponding cortical representation, and this has consequences for the perception of touch. When, for example, two touch stimuli are simultaneously applied to the index finger, they are consistently misjudged as being farther apart than when they are applied to the less sensitive forearm or lower back, even though their distance apart is identical in all three instances. This "rescaling" can be eliminated by distorting the perceived size of the body parts—when the forearm is seen at twice its normal size, and the hand at half its normal size for a short period of time, the same two touch stimuli are now perceived as being the same distance apart when applied to each of those body parts. Body posture can also influence the perception of touch. When two touch stimuli are applied to each hand, one after the other, their timing and location are harder to judge when the arms are crossed than when they are not.

To account for such findings, some researchers have subdivided the body schema into three separate components: the primary somatosensory representation, a secondary "body form" representation of body size and shape, and a third "postural representation" of body part position, which also encodes the location of touch applied to the body. Thus perception of touch occurs as a two-stage process, so that a stimulus is mapped first onto the primary somatosensory representation and then "transformed" and mapped onto the postural representation. Initial processing of touch on the hand would therefore assume that the arms are not crossed, and body posture would only be taken into account during the second stage of processing.[39]

Two London researchers hypothesized something similar in 2010. Matthew Longo and Patrick Haggard sat their participants at a table with their left hand under a board in front of them and asked them to judge the position of each fingertip and knuckle by moving a baton on the board. By averaging the distances between the "judged" positions of fingertips and knuckles from dozens of attempts by each participant, they showed that all of the participants consistently perceived their hands to be massively distorted, like Penfield's homunculus, with shorter fingers and a wider overall hand size. Curiously, though, these distortions appear to occur only for the back of the hand—participants' estimates of hand size were far more accurate when their palms faced upward. This is transferred to the perception of objects, too: objects feel larger when placed across the width of the hand than when placed across the length, but only when they are placed on the back of the hand, and not on the palm. This makes sense because the palm of the hand contains a higher density of nerve endings and is therefore more sensitive than the back of the hand. People also consistently perceive less sensitive body parts to be proportionally longer than more sensitive ones; they overestimate the length of a given body part when comparing it to the length of their hand, but not when comparing it to the length of a stick, and also underestimate the length of the body part when comparing it to the length of the less sensitive forearm. From this, the researchers conclude that the brain encodes a "body model"—an implicit representation of the size and shape of the body—which is distinct from both the body schema and the body image.[40]

The more recent "body model theory" proposes that the somatosensory cortex does not function simply as a sensory map of the body. According to this theory, different layers of the somatosensory cortex cooperate to perform two distinct body-related tasks. Microcircuits in layer 4 of the cortex generate a model of the body, based on genetic information and sensory inputs. The layers above and below store sensory and motor memories, respectively, and feed into layer 4, via layer 6, to animate it. Thus the somatosensory cortex functions more like a flight simulator than a map: It not only generates a representation of the body, but it also runs body simulations that can be used to mentally rehearse and evaluate actions.[41]

Other brain regions also contain body maps. The visual system, which occupies a sizable portion of the occipital lobe at the back of the brain, as well as parts of the temporal lobe, contains several distinct areas that process body-related information. Visual processing takes place in stages, increasing in complexity with each stage. Light information from the retina is transmitted to the very back of the brain, which contains cells that respond to the most basic features of a visual scene, such as contrast and the orientation of edges. As a general rule, each successive stage processes increasingly complex visual information and takes place progressively farther away from the back of the brain. Thus the lower surface of the temporal lobe contains an area that is specialized for processing visual images of human faces (the "fusiform face area"), and this overlaps with, but is distinct from, an adjacent area (the "fusiform body area") involved in processing visual images of human bodies. A more recently identified region, lying at the border of the occipital and temporal lobes, contains a highly organized body map that is selectively activated in response to an individual's viewing images of their specific body parts, such as upper and lower limbs, and to the individual's own movements of these same body parts. This body map is organized topographically, like Penfield's homunculus, with overrepresentation of the upper limbs; but, unlike the homunculus, it does not overrepresent the face. The brain region containing this map also appears to categorize body parts into those which can perform actions (such as the hands and feet) and those which cannot (the face and chest), and to be involved in manipulating mental images of these body parts.[42]

Findings from the few studies investigating phantom limbs in congenital amputees complicate matters even further. It was once widely believed that children born without one or more limbs and those who have had a limb amputated in early life do not experience phantom limbs because the missing limbs are not represented in the body schema or body image. Research published in the 1960s shows that this is not the case, however. In the first such study, in 1964, researchers interviewed more than 100 children born missing one or more limbs or parts of limbs, many of whom described phantom sensations very similar to those reported by adult amputees. The children described vivid sensations of phantom limbs that felt "real" and had a distinct shape and position and could be moved voluntarily. This study noted that phantom limb sensations tend to be more common in children missing a hand or foot than in those missing intermediate sections of a limb, a finding which, according to the researchers, is "consistent with greater cortical representation of such parts." A second early study, also in 1964, described the case of an eleven-year-old girl who was born missing both forearms and hands, who "reported very distinct and intense phantoms which she had experienced for the first time at the age of six years" and who thereafter "had the feeling of two completely normal hands, placed about 15 cms [6 inches] below the stumps." She was, the study continued, "able to differentiate and to move freely all of her fingers . . . [and] in school she had learned to solve simple arithmetic problems by counting with her fingers just as other healthy children did [by placing] her phantom hands on the table [to] count the outstretched fingers one by one."[43]

Modern research suggests that at least one-fifth of congenital amputees experience phantom limbs, and this is taken as evidence for a "hard-wired" body representation that is at least partly determined by genetics. On the other hand, only about half of young children who undergo amputation before age six experience phantom limbs—far fewer than those who undergo amputation as adults—which could be because the children did not receive sufficient sensory experience to incorporate the missing limb into a "live" body image whose development relies on sensory information. Thus, although some body representations seem to be highly dynamic and can be altered in response to experience, others seem to be "hard wired." And so, as well as distinguishing between explicit and implicit body representations, some researchers also distinguish between "online" representations, which are updated in real time by sensory information, and "offline" representations, which are not. According to one hypothesis, proposed by Manos Tsakiris as a "neurocognitive model of body ownership" in 2010, the sense of body ownership may arise by combining online—"live"—multisensory information with offline information about the body.[44] A "body image"–type representation is likely to incorporate information from all such "lower-order" representations. We become aware of our body through multisensory integration: different types of sensory signals are combined in our brain to generate our conscious experience of having a body and being in control of it. It is now widely believed that the integration of visual (sight), tactile (touch), and proprioceptive (perception of bodily posture, movement, and balance) information is critical for bodily awareness, and that bodily self-consciousness is an ongoing process that is largely dependent on a continuous influx of these multisensory streams from the body into the brain. Altering the body-related information that enters our brain can profoundly alter the way we perceive and experience our body; and it is becoming increasingly clear that other sensory modalities, such as hearing, also have a role to play in bodily awareness.

5 Multisensory Integration

> We have tactile, thermal, pain impressions. There are sensations which come from the muscles . . . and sensations coming from the viscera. Beyond that there is the immediate experience that there is a unity of the body. This unity is perceived, yet it is more than a perception.
>
> —Paul Schilder, *The Image and Appearance of the Human Body*

Bodily awareness is based on a continuous stream of several different types of sensory information, including information from our "traditional" senses, which usually gives rise to our conscious perceptions, and information from our other senses, which enters our brain but does not enter our conscious awareness. Thus we see our body from a certain perspective and hear the various sounds that it makes. When we move, we may become aware of touch sensations generated by our movements; at the same time, stretch and pressure sensors in our joints and muscles send signals to our brain. All of this information is combined in our brain, which uses it to update its body representations and to guide our actions. The flow of sensory information into the brain can be interrupted or manipulated easily, and researchers have exploited this to create various illusions of bodily awareness—illusions that can dramatically alter the way we perceive our body, and thus also our self.

The idea that we have five senses is usually attributed to Aristotle. The Greek philosopher was an empiricist, who believed that "all knowledge begins with the senses." In *De Anima* (*On the Soul*), published around 350 BCE, he devoted one chapter each to sight, hearing, smell, taste, and touch, and he argued that there are no other senses than these five. But

Aristotle was wrong: there are more than five senses. Moreover, the senses do not operate independently: our perceptions are multisensory, involving sensations from two or more sensory modalities, which cooperate to enhance one another and enrich our subjective conscious experiences. This is true for most everyday experiences. We can perceive speech with our ears alone, but looking at a speaker's mouth enhances our perception of speech, and this is especially useful when we are trying to hear the speaker's words over background noise. Eating is also a multisensory experience. Our ability to taste food depends largely on our sense of smell, but the sight, sound, and texture of what we eat also influences how we perceive its flavor.

Sight

Bodily awareness depends on multisensory integration, too, with our sense of sight being vitally important. We normally experience our self as being located within the physical boundaries of our body, and usually view our own body from the first-person perspective. This subjective point of view is disturbed in out-of-body experiences, during which the mechanisms of self-localization break down, such that our "self" becomes disembodied. We feel that we are located elsewhere, and that we are viewing our body from a third-person perspective, or from outside.

Although out-of-body experiences can occur following a stroke, during epileptic seizures, or as an effect of certain drugs, Bigna Lenggenhager and colleagues demonstrated in a 2007 study that these experiences can also be induced in people with no brain damage under laboratory conditions. In a variation of the rubber hand illusion, study participants wore a head-mounted visual display connected to a video camera placed several meters behind them, and thus saw images of their own body, viewed from the back, in real time. The experimenter would stand beside a participant and, using two plastic rods, repeatedly stroke and prod the participant's body and the video image of it. Touching the participant's real and "virtual" body at the same time created the illusion that the participant was standing several meters behind his or her body and viewing it from that position (figure 5.1). Immediately afterward, participants reported feeling as though they were outside their body looking at themselves from behind. And when blindfolded, moved, and asked to return to their original location, they drifted toward the position of their virtual body. The conflict between their sense of sight and their sense of touch caused them to localize their "self" outside their physical body, but only when touch on their real body was synchronous with that on their virtual body. Following the illusion, participants also exhibited a reduced fear response when they saw their own body being threatened with a knife, suggesting that they had, at least to some extent, "disowned" their real body and transferred their sense of body ownership to the virtual body.[1]

Manipulating the visual perspective of the body can create various other strange experiences. In the "body swap illusion," described in a 2008 study by H. Henrik Ehrsson and his colleague Velena Petkova, the video camera is attached to a helmet worn by a life-size mannequin and points downward over its body, so that a participant wearing the head-mounted display sees only the mannequin's body, which, through synchronous touch, the participant then perceives as their own. A full-blown body swap illusion can be created when the helmet is worn by another person. When the two persons stand opposite each other, the participant sees their own body in the head-mounted display, viewed from the other person's perspective; the participant and the other person then hold and repeatedly squeeze each other's hands synchronously for several minutes; which creates in the participant's mind the illusion of being "inside" the other person's body, even though it also appears that they are shaking hands with themselves. The illusion worked with a mannequin and a person of the opposite sex, but broke down when the two squeezed each other's hands at different times, and it did not work at all when the camera was placed on a nonhumanoid object, such as a large box of the same size as the mannequin's body.[2]

Another version of the body swap illusion is the "Barbie doll illusion," in which the head-mounted display is connected to a video camera that looks down over the body of a small doll. In this scenario, synchronous stroking creates in the participant the experience of taking body ownership of the doll's body and also alters the participant's perception of size and distance of nearby objects, which appear to be far larger and farther away than they actually are. Conversely, taking ownership of an oversized artificial body makes the same objects seem smaller and closer to the participant.[3]

Similar experiments show that study participants will also take ownership of a computer-generated full body or body part within virtual reality environments. In "virtual hand illusion" experiments, participants sit in front of a back-projecting screen displaying a three-dimensional arm that appears to be attached to their shoulder. Tapping their arm with a motion-tracking wand that replicates the movements to the screen creates the powerful sense that the touch sensations come from the arm on the screen, but again, this illusion breaks down when the wand touches the real and the virtual arm at different times. And, as a 2012 study showed, the same method can be used to induce ownership of a virtual arm that is twice, or even three times, as long as the real arm, although this "very long arm illusion" becomes weaker the longer the arm.[4]

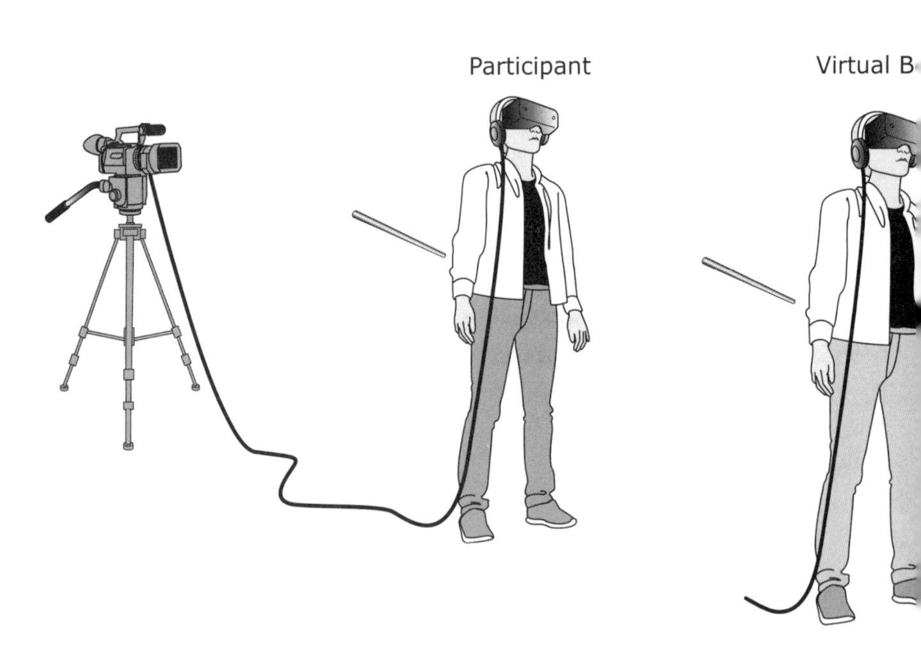

Figure 5.1
The virtual body illusion. The participant stands in front of a video camera wearing a head-mounted visual display that shows footage from the camera, so that they see their own body projected in front of them. The experimenter then prods the participant's back with a stick. When the camera shows live footage, the participant sees and feels the prodding synchronously, and begins to take ownership of the virtual image of their body, or experience it as if it were their real body. When the display shows recorded footage, the touch sensations felt and seen by the participant do not occur synchronously, and they do not experience the illusion. Adapted from Lenggenhager, B., Tadi, T., Metzinger, T., and Blanke, O. (2007), Video ergo sum: Manipulating bodily self-consciousness, *Science*, 317:1098.

Just as the rubber hand illusion can be extended into the full body illusion, so, too, can the virtual hand illusion be extended such that participants can take ownership of a whole computer-generated body, including one of the opposite sex.[5]

Multisensory illusions that alter body perception can influence the way we think. White people who take ownership of a dark-skinned rubber hand or of a Black virtual body exhibit a sustained reduction in implicit racial bias afterward. One of the most recent multisensory illusions is the "friend–body swap illusion," in which researchers in a 2020 study used head-mounted displays to induce the body swap illusion in pairs of friends, after they had filled out questionnaires to rate each other's personality traits. During the illusion, they rated their own personality traits as being more similar to those of their friend, and this was related to the strength of the illusion, with greater similarities for those who experienced it more robustly. Furthermore, participants had better recall of the personality traits they rated during the illusion than those they rated during the no-illusion sessions of the experiment, suggesting that a close match between body perception and self-concept may be necessary for proper memory encoding.[6]

A 2019 study showed that bodily self-consciousness plays a role in memory function. Episodic memories, or memories of everyday events, typically get weaker with time, but they can be enhanced when the body is viewed in a virtual scene during memory formation. Study participants wore head-mounted displays showing immersive virtual environments consisting of digitized 360-degree video recordings of scenes from everyday life. They entered and explored three different rooms and tracked a virtual ball programmed to appear and move through each room. Some scenes included the participants' own body viewed from a first-person perspective, and in these they pointed at the ball and tracked its movements. In other scenes, their body was not present or was replaced with an object. Later, when participants were tested, they recalled more details of scenes that included their body than of scenes that did not.[7]

The ability to transfer the sense of body ownership could prove to be useful for telerobotics, the control of semiautonomous robots from a distance. Researchers in Japan have transferred body ownership to a humanoid robot. In a 2012 and a 2013 study, participant-operators wore a motion-capture system that transferred their grasping movements to the robot's hands; their intended movements matched the visual feedback from the robot's body and the sensory feedback from their own movements, creating the illusion that they had taken ownership of the robot's body. The illusion persisted even when there was a one-second delay between their movements and those of the robot, although it was slightly weaker. The operators could also control the robot's body without moving a muscle, using scalp electrodes that translated their movement-associated brain waves into the robot's hand motions. This again highlights the importance of the sense of sight in bodily awareness—visual feedback alone, combined with the operators' imagined movements, was sufficient to induce a feeling of ownership over the robot's body. These findings could eventually enable teleoperators to take ownership of—and thus gain better control of—robots roaming the seafloor or the surface of distant planets. But this may only be possible if the robots, or at least parts of them, have a humanlike appearance.[8]

Simple manipulations of the appearance of the body can alleviate phantom limb pain in amputees. "Mirror box therapy" requires a basic apparatus consisting of a large rectangular box with its front and top sides removed, containing a vertical mirror attached to one side. Thus one-handed amputees are asked to sit in front of the box and to place their present hand in the box so that its reflection appears where their missing hand would be. They then move their present hand while looking into the mirror, creating the illusion that both hands are performing identical, symmetrical movements.

Mirror box therapy was first attempted in 1993, with a male amputee patient known as "D.S.," who had his left arm amputated above the elbow eleven years earlier and had experienced continuous, excruciating phantom pain ever since. D.S. took a mirror box home and used it for ten minutes per day for two weeks, during which time he reported that his paralyzed phantom arm began to move, and the pain he felt was significantly reduced. After another week of mirror box therapy, he reported that his phantom arm had disappeared altogether, along with the pain in the elbow and forearm, but that his phantom fingers now dangled from his shoulder, and were still painful. Amputees like D.S. often report experiencing extremely painful phantom sensations in the days and weeks after amputation, in which they feel their missing hand clenching into a fist with their nails digging into the palm; the mirror box procedure can immediately relieve some new amputees of these clenching spasms, too. Exactly why such a simple procedure can alleviate phantom pain is still unclear. One hypothesis for why amputees experience phantom pain is that the brain does not receive sensory feedback from the missing limb, and so the mirror box may work by creating the visual illusion that the missing limb is in fact present.[9]

The effectiveness of the mirror box relies on patients paying constant attention to the reflected present limb, as the illusion breaks down if they look away. Furthermore, the box restricts the movements of the present limb, and requires patients to keep their whole body in the same place. In a 2006 study, researchers in Manchester, England, used virtual reality to overcome these limitations. They have developed a system consisting of a head-mounted display that shows images of an arm amputee's whole body within a virtual reality environment, operated by a "data glove" fitted with motion sensors and worn on the remaining arm. The system software includes an algorithm that transposes movements of the real limb to the virtual limb. This allows the patient to move and act freely without compromising the illusion of the missing limb. It can also be used by lower-limb amputees by means of sensors attached the knee and ankle joints of the present leg.[10]

Early virtual and augmented reality systems developed in the 1960s created simulated environments for military and medical training purposes, and these are still the two main applications of the technology. The other is in the gaming industry, and there are now a number of affordable commercial devices available for use with both consoles and smartphones. State-of-the-art virtual reality systems deliver images to both eyes through mobile headsets to re-create binocular stereoscopic vision, combined with surround sound to produce an immersive experience. Most use computer-generated images, which can be delivered to a head-mounted apparatus using Wi-Fi, but some systems now use video footage of real images. Modern systems employ motion-tracking technology to digitize the user's movements within the virtual environment, and some of the more sophisticated systems are now also incorporating eye-tracking technology.

Next-generation virtual reality systems already being developed add the dimension of touch. Vibrating controllers for arcade games and consoles are commonplace, and even though haptic gloves, shoes, and full-body suits are already available, they are not yet widely used in virtual reality systems, perhaps because they are still somewhat cumbersome. Various research groups are, however, exploiting rapid advances in materials science and electronics in the development of soft artificial skin interfaces that communicate mechanical vibrations and other types of touch information to the body. Eventually, such devices will not only transform the virtual reality experience and the gaming industry; they will also be incorporated into the next generation of artificial limbs and may also be used to enhance social media interactions.[11]

Proprioception and Touch

> The main sensual basis of the awareness of the limb and its posture rests, as disease and experiments show, on the sensory nerves of the motor structures in the limb [which] . . . collectively register the tension at thousands of points they sample in the muscles, tendons, and ligaments of the limb. These keep on firing into the nerve-centers nerve-impulses registering, so to say, the tension of sample-points within the limb. In my awareness of the limb and its posture, and similarly in my awareness of its movements when it moves, I perceive no trace of all this. In "experiencing" the limb I find no hint of this multiplex origin of the percept, no additive character in it, no tale of tensions within the limb, or of its possessing muscles or tendons. I am simply aware of where the limb is, and when it moves—or is moved, for my moving it myself hardly helps my perception of it further. The percept is not a very vivid one.
>
> —Charles Sherrington, *Man on His Nature*

The rubber hand illusion manipulates visual and tactile information to create a discrepancy between what we see and what we feel, to create the feeling that a fake hand is actually real and our own. Earlier experiments manipulated proprioceptive information to create illusory limb movements.

Position sense—proprioception—arises from stretch receptors in the joints, muscles, and tendons, which are activated by the tiniest of movements, sending signals to the brain. But activating these sensors artificially produces the illusion that a limb is moving when it is not, and even that it is moving into impossible positions. For example, if the wrist is extended up to its anatomical limit and the wrist flexor muscle in the forearm is vibrated with a mechanical buzzer, the hand feels as if it is bent farther upward, at a right angle to the arm. Applying the buzzer to the biceps muscle in the arm creates the illusion that the arm is first straightening and then bending beyond its limit, causing study subjects to report that "my arm is broken" or that "it is being bent backwards," whereas vibrating the triceps creates illusory movements in the opposite direction, so that a study subject may report that "my hand is going through my shoulder." This creates a great deal of confusion about the actual position of the arm, and, indeed, sometimes produces the illusory feeling of having multiple forearms. The illusory movements are perceived as if they are real—subjects typically adjust their gaze to locate the illusory limb in its new position, and if a light is attached to it, they see the limb move in the direction of the illusory motion, even though their real arm has remained stationary.[12]

Likewise, vibrating the biceps while subjects are touching their nose with it creates the "Pinocchio illusion"—the perception that their nose is growing longer and longer, in keeping with the illusory movement of their forearm, which is perceived to move farther and farther away from their face. The procedure can also be used to distort the perceived size of other body parts. If subjects stand up straight with both arms akimbo, vibrations applied to both of their biceps elicit an illusory extension of both their forearms, along with the sensation that their waist is expanding at the point where it comes into contact with their hands. Conversely, vibrations of the triceps muscles of both arms induce the illusory sensation that their forearms are flexing, and that their waist is shrinking accordingly. Thus vibrations of the right frequency and in the right place trick the stretch receptors into sending signals of illusory changes to the brain, which then updates its online body representations accordingly, creating illusory changes in how we perceive the orientation, configuration, and dimensions of our body.[13]

Proprioception, the sense of the position and movement of our body and of our limbs, joints, and other body parts, was first described in the 1830s by the Scottish anatomist-surgeon Charles Bell, who thought of it as a "sixth sense": "The muscles are from habit so directed with so much precision that we do not know how we stand . . . [but] if we attempt to walk on a narrow ledge we become subject to apprehension . . . the actions of the muscles are magnified and demonstrative."[14]

In humans and other mammals, proprioceptive organs are found in the skeletal muscles, tendons, and joints. Within our muscle fibers are capsules of connective tissue called muscle spindles, which contain specialized muscle fibers that run parallel to the main fibers that contract to produce skeletal movements. Whereas the main muscle fibers are innervated (supplied with nerves) by large motor neurons that send the signals from the brain to initiate these contractions, fibers inside the muscle spindles are innervated by sensory neurons, whose endings spiral around the central portion of the spindle, and which are sensitive to both the length of the muscle fibers and the rate of change in their length. Their large-diameter insulated fibers send fast signals about these muscle properties up to the brain via the spinal cord. Spindle edges are innervated by another type of fast sensory neuron that is sensitive only to muscle spindle length. The tendons, which attach muscles to bones, contain Golgi organs, each of which contains the nerve endings of a single sensory neuron of yet another type, which wraps around strands of collagen fibers attached to an individual muscle fiber. The sensory neurons of this type send slightly slower signals about muscle tension during movement, becoming more active as muscle tension rises. Finally, the joints contain organs called Ruffini endings and Pacinian corpuscles, each of which transmits signals when its joint reaches a certain threshold or position.[15]*

Whereas the visual pathway transmits light information from the eyes to the brain, the auditory pathway transmits sound information from the ears to the brain, and so on, the somatosensory pathway transmits not only proprioceptive information from the muscles, joints, and tendons, but also multiple, distinct types of touch and pain signals from the skin to the brain. Through it, we can determine the shape and texture of objects and discriminate between a cold beer and a hot stove, a dull ache and a prickly pain, a loving caress and an angry prod.

We are only just beginning to grasp the complexity of the body's somatosensory system. All bodily sensations and proprioceptive signals are transmitted to the brain by primary sensory neurons, each of which has a single fiber emanating from a cell body located just outside the back of the spinal cord. This fiber splits into two, with one end projecting out to the skin, muscles, and tendons, and the other projecting into the spinal cord. Distinct subsets of primary sensory neurons transmit specific types of information. Some are dedicated to dull, burning pain and some to sharp, stabbing pain; some are dedicated to cold temperatures and others to hot. There are also subsets that are dedicated to itch and to pleasant, "affective"

* That proprioception is critical for coordinating movement is evidenced by the fact that nearly all animals capable of movement have this sense. Indeed, entomologists had described the proprioceptive organs of insects by the mid-nineteenth century, although the precise function of these organs remained unclear for a few decades. Even the simplest of all animals, such as the 1-millimeter-long nematode worm *Caenorhabditis elegans*, whose entire nervous system contains just 302 neurons, has proprioceptors that send feedback to the neural circuit controlling movement. Land plants also use proprioception to sense their position and to maintain an upright posture.

touch, and it may be the case that there are still other subsets dedicated to other sensations such as tickling.

Primary sensory neurons have the same basic structure but differ widely in certain other, important respects, namely, in the diameter of their fiber, in the extent to which they are insulated by myelin, in the protein sensors present at the peripheral nerve ending, and in exactly where the centrally emanating end of the fiber terminates. The largest-diameter fibers, called A-alpha fibers, are thickly insulated by myelin and propagate nervous impulses at velocities of up to 120 meters (400 feet) per second, whereas small-diameter C fibers are uninsulated and have conduction velocities of 2 meters (7 feet) per second. A-alpha fibers carry impulses that encode proprioceptive information, whereas C fibers carry impulses that encode information about dull aching pain, cold pain, and pleasurable deep pressure. These and other fibers bundle together into the peripheral nerves, which enter the back of the spinal cord, and either connect with secondary sensory cells upon entry or climb all the way up to the brain stem. (The fibers of motor neurons, which signal muscles to contract, leave the front of the spinal cord and bundle into the same peripheral nerves.) In the spinal cord or brain stem, primary sensory neurons form connections with second-order neurons, which project their fibers into a brain structure called the thalamus (meaning "deep chamber"), where they in turn connect to third-order neurons whose fibers terminate in the primary somatosensory cortex. Although the thalamus was once thought to be no more than a relay station through which neurons in all sensory pathways pass on their way to corresponding regions of the cerebral cortex, we now know that it also contains somatotopic maps and plays important roles in the processing of sensory information. Indeed, conscious awareness of all types of sensory stimuli involves both the thalamus and the cerebral cortex, and the two structures are connected by multiple reciprocating circuits. The thalamus thus acts as a "gateway" for sensory inputs to the cortex—and, perhaps, to conscious awareness itself.

From the spinal cord, primary sensory neurons extend fibers out to the skin, where their endings branch out to form a receptive field and express specific combinations of protein sensors that determine the type and range of stimuli they will respond to within that field. Chief among these proteins are molecules called transient receptor potential channels, more simply known as TRP channels, and Piezo channels, which span the nerve terminal cell membrane to form pores that allow electricity, in the form of positively charged ions, to flow into the cell when they are activated. This initiates nervous impulses that travel up into the spinal cord.

TRP channels are a family of proteins that are sensitive to various types of noxious stimuli capable of causing harm or injury, such as extreme heat, cold, or pressure. Individual molecules are activated over a small temperature range and by mechanical pressure. They also contain binding sites to various chemicals, some of which occur naturally in the outside world, and some of which are released from damaged tissues. For example, TRPV1 and TRPV2 are activated by temperatures higher than 42°C (108°F) and 52°C (126°F), respectively, and both also contain a binding site for capsaicin, the active ingredient of chili peppers. Thus it is because of their activation that we feel a burning sensation when we touch a painfully hot object or eat spicy food. At the other end of the scale, TRPM8 is sensitive to ambient temperatures of below about 26°C (79°F) and also contains binding sites for cooling chemicals such as mint and menthol. Between them, the eight different TRP channels found in primary sensory neurons respond to a wide range of hot and cold temperatures, and through their mechanical sensitivity they also play a role in detecting other types of pain sensations, as well as touch and itch. Indeed, the multiple roles that TRP channels play in pain perception make them prime targets for the development of new analgesics.[16]*

Piezo channels are pressure sensors found in subsets of primary sensory neurons that signal light touch, and in the proprioceptive neurons that innervate the muscle spindles and Golgi tendons. Their discovery began with the identification of mutations that inactivated the genes encoding the Piezo proteins and synthesize nonfunctional molecules in two young patients with "undiagnosed neuromuscular disorders." These patients presented with touch deficits along with neuromuscular problems and skeletal deformities, including a progressive and severe sideways curvature of the spine. Neither of them began to walk until between six and seven years of age; both now walk unsteadily—but not with their eyes closed—and have difficulty dressing and feeding themselves. They have greatly reduced

* Because of the temperature sensitivity of TRP channels, some snake species—boas, pythons, and pit vipers—have adapted them for use in thermal vision to hunt prey in the dark.

sensitivity to light touch on their hands and fingertips and are completely insensitive to vibration on hairy areas of their skin. But they respond normally to gentle stroking of these areas, and they presented with no deficits in their senses of pain, itch, and temperature.[17]

In his 1995 book *Pride and a Daily Marathon*, the neurologist Jonathan Cole describes the case of Ian Waterman, who at nineteen years of age experienced a rare autoimmune response to the influenza virus, which attacked the sensory neurons from his neck down, blocking all touch and proprioceptive signals from his entire body:

> He could feel nothing from the neck. Nor could he feel his mouth and tongue. Not only couldn't he feel anything to touch, he had no idea of where the various bits of his body were without looking at them. He could not feel anything with his arms, his legs or his body. That was frightening enough, but he had no awareness of their position either. It wasn't that the muscular power was affected, since he could make an arm move. But he had no ability to control the speed or direction of the movement. Any movement happened in a totally unexpected way. It was pointless to try.[18]

Conditions like Waterman's are extremely rare and did not come to the attention of clinicians until the 1990s. But once they had been identified, geneticists were able to isolate the two genes that encode human Piezo channels, and to characterize the structure and possible function of the channels themselves at the molecular level. Individual Piezo channels are made up of three smaller, curved molecules that are arranged like propeller blades around a central pore, forming a domed structure over the opening of the pore, which juts out from the outer surface of the cell membrane. This arrangement suggests an elegant mechanism by which mechanical force is transformed into electrical impulses: mechanical pressure upon the membrane pushes the dome down into the membrane, causing the propeller blades to rotate and the pore to open.[19]

Hearing

Thus far, research into the multisensory basis of bodily awareness has focused exclusively on the respective roles of sight, touch, and proprioception. The potential role of hearing remains largely unexplored, but recent findings suggest that it, too, plays a role in shaping our representations and perception of the body.

Manipulating the natural sounds our body makes when it comes into contact with surfaces and objects can alter the way we perceive certain properties of our body. For example, several studies show that altering the sounds associated with touch can alter the perception of felt sensations. In a 1998 study, participants rubbed their hands together while a nearby device recorded the sounds made by their movements and played them back in real time through headphones. The auditory feedback was either identical to the original sound, or altered so that the high frequencies were boosted or diminished. Changing the loudness of these frequencies altered how the participants perceived their skin texture—they described it as smoother and drier when the high frequencies in the sounds they heard were boosted, and rougher and wetter when those frequencies were diminished. In a 2014 study, participants' right hands, hidden from their view, were tapped with a small hammer while the natural sound of the tapping was replaced with that of a hammer hitting a piece of marble. After a few minutes, the participants started to feel as if their right hand was actually made of marble—it began to feel stiffer, harder, and heavier.[20]

More recent research shows that unnatural sounds can also alter perception of the body. In a 2017 auditory version of the "Pinocchio illusion," participants sat at a table with both their hands hidden behind a black sheet. They used their left hand to pull on their right index fingertip while they listened to sounds of rising, falling, or constant pitch through headphones. They felt and estimated their fingers to be longer immediately after hearing sounds with a higher pitch. Likewise, as a 2018 study showed, manipulating the auditory and tactile cues from falling objects can alter participants' perception of how tall they are. Researchers did so using an apparatus that consisted of a metal frame fitted with a nylon motion-sensitive net and suspended over a loudspeaker and a mechanical shaker attached to a rubber mat. Blindfolded study participants stood over this apparatus and dropped balls from head height, which were caught in the net and triggered the loudspeaker and shaker to mimic the balls hitting the floor at predetermined time delays corresponding to different heights. With longer delays, the balls seemed to hit the floor later than expected, causing the participants to report that they felt taller than they actually were. They also behaved as if their legs were longer—afterward, when asked to move back to a position they had memorized, they took shorter steps. Shorter time delays had the opposite effect, making the participants feel shorter than they actually were and take longer steps toward the memorized position.[21]

The way we walk and the sounds we make with our footsteps convey a surprisingly large amount of information about us. Walking sounds differ depending on the properties of the shoes we are wearing and the type of surface we are walking on, and other people can accurately judge not only our pace and posture, but also our gender, and even our emotional state, from the timing and frequency of these walking sounds. As well as conveying information to others, walking sounds are also an important source of feedback for the person making them, and one 2015 study shows that manipulating these sounds can alter the perception of certain characteristics of the body. Participants wore sandals equipped with microphones and force-sensitive resistors, attached to an amplifier and equalizer, which sensed their footsteps, changed the sounds they produced, and relayed the altered sounds to them in real time through headphones. Hearing these altered sounds in time with their footsteps not only changed the way the participants perceived their body weight, but also altered their feelings about their body and caused them to change the way they walked. Amplifying the high frequency components of the walking sounds produced sounds consistent with a lighter body. This made the participants feel as if their body actually was slightly lighter, and seemed also to change their attitude toward their body and its capabilities, so that their gait became slightly faster and more upbeat. Hearing walking sounds with the low frequencies amplified had the opposite effect. The prototype sandals could quite easily be developed into an apparatus that alters how people who are overweight or who have eating disorders such as anorexia nervosa perceive their body.[22]

All information from our senses is transmitted from our sense organs to the particular sense's corresponding cortical area in the brain: visual information to the primary visual cortex in the occipital lobe at the very back of the brain, auditory information to the primary auditory cortex in the temporal lobe, and tactile and proprioceptive information to their respective areas in the primary somatosensory cortex. All of these brain areas are referred to as "primary areas" because they are the first port of call in the cerebral cortex for a given type of sensory information. As Wilder Penfield observed almost a century ago, however, some parts of the somatosensory cortex respond to movement, and some parts of the adjacent motor cortex respond to touch. In fact, all of the primary sensory cortical areas are interconnected; the area for each sense likely contains "multimodal cells" that respond to and begin to integrate information from the other senses. The properties of individual cells in the primary sensory areas are highly adaptable and can be dramatically modified in response to experience. Thus the primary visual cortex in people who are blind can process auditory information and, for those of them who learn to read Braille, tactile information, too.

Beginning in the primary sensory cortical areas, processing continues in the "association areas" located in the temporal and parietal lobes of the brain, which perform the bulk of the multisensory integration that underlies our sense of bodily awareness. Some brain scanning studies show that the illusory sense of limb ownership felt during the rubber hand illusion is associated with activity in multiple regions of the human brain, including the somatosensory and motor cortices and parts of the parietal lobe, and other studies show that these corresponding regions of the monkey brain contain individual "trimodal" neurons that respond to and integrate visual, tactile, and proprioceptive information. Similarly, activity in a region lying at the junction of the temporal and parietal lobes of the human brain is associated with participants feeling located within their own body and seeing their body from the first-person perspective, whereas diminished or extinguished activity in this same region when it is damaged is associated with out-of-body experiences.[23]

The disturbances of bodily awareness described by Henry Head and Paul Schilder were most often associated with damage to the parietal lobes of the brain, but such disturbances can occur as a result of damage to numerous other parts of the nervous system. Thus somatosensory processing can be disrupted not only by damage to the primary somatosensory cortex, but also (1) by mutations affecting the function of the various touch and pressure sensors in nerve endings; (2) by damage to the peripheral nerves that transmit sensory information into the spinal cord; (3) by damage to the spinal cord itself, which contains the ascending tracts that transmit multisensory information into the brain; and (4) by damage to the thalamus, near the center of the brain. Damage to any of these parts of the nervous system would lead to "*dis*integration" of the multisensory processes underlying bodily awareness and thus also to altered perception of the body.

Development of Multisensory Integration

Several research groups are investigating the development of multisensory integration by performing the rubber hand illusion on children. The work in a 2013 study, for example, shows that children aged four to nine years are also susceptible to the illusion; after induction of the illusion, however, children across this age range perceive the position of their real hand to be closer to the fake hand than do adults. This suggests that they rely on visual and touch information, but less so on proprioception, and that visual-tactile integration begins early and remains stable throughout early childhood. But, as a 2016 study shows, slightly older children perform similarly to adults, leading the researchers to conclude that multisensory integration underlying bodily awareness undergoes a prolonged period of development and reaches maturity by ten to eleven years of age.[24]

Recent technological advances enable researchers to scan the brains of infants; indeed, some researchers are now using these methods to investigate the development of bodily awareness during the first months of life. In infants as young as five months of age, light touch applied to the hands and feet elicits activity in the sensorimotor regions that corresponds to somatotopic organization, suggesting that online body maps are already present, or at least are starting to develop. These same brain regions are also activated not only when infants move their own body parts, but also when they observe others performing the same movements. The somatosensory system also distinguishes between self-touch and the touch and others, and is activated in response to seeing others being touched or in pain. Body representations thus seem to be crucial for social perception and cognition—they appear to play a role in distinguishing self from other, in learning by imitation, and in our ability to empathize.[25] Further research into the social role of body representations could improve our understanding of the origins of neurodevelopmental conditions such as autism spectrum disorders, which are characterized by impairments in communication and social interactions.

Our new understanding of how bodily awareness develops also allows us to speculate on the origins of body integrity identity disorder (BIID). One study, published in 2013, used structural magnetic resonance imaging to examine and compare brain structure in thirteen male BIID subjects and thirteen male non-BIID controls of the same age. In their analysis of the brain scanning data, the researchers used specialized computer software to reconstruct the architecture of the cerebral cortex, looking at measures such as surface area and cortical thickness. Averaging the results from all twenty-six study subjects, they found that the cortex in the right superior parietal lobule was markedly thinner in BIID subjects than in the controls. Furthermore, they also noticed a reduction in the surface area of BIID subjects' right inferior parietal lobule and found a negative inverse correlation between the extent of this reduction and the magnitude of the desire for amputation; that is, the more a BIID subject wanted a limb to be amputated, the greater was the observed reduction in surface area of the right inferior parietal lobule. These results are consistent with the hypothesis that the desire to have a limb amputated arises when the higher-order neural representation of that limb is weaker than it should be.[26]

More recent studies show that subjects with BIID also exhibit significantly reduced activation of their right parietal lobe in response to touch stimuli applied to the limb they wish to have amputated, compared with their response to touch on their other, opposite limb and with the response of non-BIID controls. This reduced activation is observed specifically in the superior parietal lobule, a brain region that is suspected to integrate different types of sensory inputs to generate a higher-order representation of the body. The latest such study, published in 2020, further shows that sensorimotor brain regions are less well connected to other brain regions in BIID subjects than in control participants, and that gray matter volume in both the right superior parietal lobule and the premotor area, which is involved in planning movements and which may also integrate sensory information from the limbs, is reduced. Furthermore, gray matter volume in the parietal lobe of BIID subjects appears to be related to the intensity of symptoms, with those who exhibit a greater reduction in gray matter volume having a stronger desire to amputate. Thus the body image of BIID subjects appears to be distorted, with the representation of the unwanted limb being diminished, or perhaps absent altogether. As a result, the subjects grow up lacking a sense of ownership over the limb they eventually want to be rid of.[27] If this is the case, we might predict that BIID subjects would be less likely to experience phantom limb sensations after actually having their unwanted limb removed.

To begin to understand how BIID might arise, we can turn to the classic experiments of David Hubel and Torsten Wiesel. Beginning in the late

1950s and using microelectrodes to investigate the architecture of the visual cortex and the properties of its cells in cats, Hubel and Wiesel made a series of important discoveries. First, they found that the visual cortex contains individual neurons that respond to the most basic properties of visual images, such as the orientation of edges. Second, they found that these orientation selective cells are organized into alternating columns, each of which preferentially receives inputs from one eye or the other. These "ocular dominance columns" are visible as a striped pattern across the surface of the visual cortex and give that part of the brain one of its other names, the "striate cortex" ("striate" meaning "striped" or "banded").

Subsequently, Hubel and Wiesel went on to investigate how visual stimulation, or lack thereof, affects development of the visual cortex. To do so, in 1963, they raised newborn kittens with one eye sewn shut, and then examined their brains. This had dramatic effects on the pattern of ocular dominance columns: the columns deprived of visual inputs failed to grow, whereas those receiving inputs from the open eye encroached on the unused areas of the striate cortex to become abnormally large. Crucially, in 1965, they found that the resulting brain changes were permanent if the kitten's eye was left shut beyond a specific stage of development, but could be reversed if the eye was reopened before that stage.[28] This work led directly to effective interventions for ambylopia, or "lazy eye," which affects up to one in fifty children, and earned Hubel and Wiesel the 1981 Nobel Prize in Physiology or Medicine. Their findings also established the importance of critical periods, or sensitive stages of development, during which the brain is both particularly sensitive to specific types of sensory inputs and dependent upon them for proper development. Now it is widely believed that all of the brain's sensory systems depend, to a greater or lesser extent, on sensory stimulation to develop properly. Certain cognitive abilities, most notably language, are also thought to develop during critical developmental periods, becoming much harder to acquire later on.

Now let us assume that body representations develop during a critical period in early to middle childhood, by an ongoing process of multisensory integration, and that, during this critical period, the brain requires a near-constant influx of sensory information from the body in order to draw up its somatotopic body maps and generate an accurate sense of bodily awareness. If the sensory inputs from one of the limbs were interrupted during this period of development for a significant amount of time, the brain would fail to generate a proper representation of the limb and would therefore not incorporate the limb into its developing body image—that is, the brain simply would not register the affected limb as being a part of the body and the self. The individual would, therefore, eventually have the strong feeling that the affected limb does not belong to them.

Anecdotal evidence supporting this hypothesis comes from the first case reports of BIID. In their 1977 paper, John Money, Russell Jobaris, and Gregg Furth described two patients who desired amputation of a healthy leg. They noted that one of the patients had spilled a pot of boiling oatmeal when he was two years old, severely burning his left leg and foot and "rendering him unable to walk for almost a year." The second patient was born with minor clubbing of his right foot, which caused it to be turned inward at the ankle, and had this deformity corrected by surgery at a young age.[29] Sensory inputs from a young child's limb might be disrupted in other ways. For example, serious injury of a limb might require it to be immobilized in a cast for a prolonged period of time, and some viruses, such as herpes viruses, infect and can damage sensory nerve fibers. Interviews with BIID patients and examination of their medical records could help to determine if, in childhood, they experienced injury or infection of the limb they wished to have amputated.

The role of the somatosensory cortex in social cognition lends itself to an explanation of how BIID patients might first become aware of their unusual desire to have an otherwise healthy arm or leg removed. Most of those interviewed to date say they first became aware of their desire for amputation after seeing an amputee at some point in their childhood. Given that an individual's observing the actions and movements of others activates representations in their own somatosensory system, the sight of an amputee could "unmask" an absent or diminished representation of a limb "to awaken an internal identity that had previously been unrecognized."[30]

Schilder did not encounter any would-be amputees, or at least he does not mention BIID in his 1935 book *The Image and Appearance of the Human Body*. He did, however, devote the final third of the book to sociological aspects of the body image, and he speculated at length about how people's body images might interact with one another.

"The body image is a social phenomenon," he wrote, and, "there is from the beginning a very close connection between the body-image of ourselves and the body-images of others." Emphasizing the link between the biological and psychological components of the body image, he argued that "the postural image . . . although it is primarily an experience of the senses, provokes [inseparable] attitudes of an emotional type . . . [such that] [w]hen we see the body of another person . . . we first get a sensory impression about [their] body . . . [which] gets real meaning by our emotional interest in the various parts of his body." Of direct relevance to BIID, Schilder stated that "optic experiences . . . lead to the construction of the body-image," and that "interest in particular parts of one's own body provokes interest in the corresponding parts of the bodies of others," and vice versa.[31]

Body representations appear to interact with one another directly in the neurological trait known as mirror-touch synesthesia. The word "synesthesia" comes from the Greek roots *syn*, meaning "joined" or "together," and *aisthesis*, meaning "perception," and refers to a group of related conditions in which stimulation of one sensory pathway simultaneously evokes sensations in another. For example, the physicist Richard Feynman described seeing mathematical equations in color, and the artist Wassily Kandinsky stated that he tried to re-create the visual equivalent of symphonies in his paintings.

Synesthesia was first described in the late nineteenth century, and though it was once thought to be extremely rare, it is now believed that up to 4 percent of the world's population experience one of its many forms. Feynman apparently had the most common form, "grapheme-color synesthesia," in which the perception of numbers or letters evokes sensations of color. Other forms include "auditory-tactile synesthesia," in which sounds elicit the perception of textures, and, more rarely, "lexical-gustatory synesthesia," in which words elicit specific tastes. In all cases, this is thought to arise from the increased growth of nerve fibers between the brain's different sensory pathways or from a failure to properly "prune" exuberant neuronal connections between them during early stages of development.

In mirror-touch synesthesia, first described in 2005, the senses of sight and touch are closely linked—specifically, when a mirror-touch synesthete sees another person being touched, they experience touch on the corresponding part of their own body. The first case report describes a forty-one-year-old woman referred to as "C.," who claimed that she had always perceived touch on other people as touch sensations on her own body, and had, until 2005, never thought this to be unusual. She also said that when she is standing opposite another person, her perceptions are mirrored, so that seeing them touch their right cheek elicits touch on her left cheek, whereas if she is standing next to another person, she experiences touch on the same side that the other person does.

Functional magnetic resonance imaging (fMRI) revealed that C.'s brain responded differently than the brain of nonsynesthetes to observing touch. In both C. and the nonsynesthetes, observing touch on the face activated the face area of the primary somatosensory cortex and a distributed network of regions thought to constitute our brain's "mirror-neuron system," containing cells that fire both when we perform goal-directed actions and when we see others performing those actions. But the levels of activity in C.'s brain were significantly greater than those seen in the brains of nonsynesthetes; furthermore, observing touch also evoked activity in C's insula, a small brain region believed to be important for self-awareness, as well as in C.'s somatosensory cortex and mirror-neuron system above the threshold for the conscious perception of touch.[32]

Mirror-touch synesthesia thus seems to involve shared representations, whereby the observed touch is transferred onto the synesthete's own body maps, leading to a mirroring or simulation of the state of the person being observed, and there is some evidence that mirror-touch synesthetes have greater difficulty than others in distinguishing their own body from that of others. The condition likely contributes to vicarious experiences such as contagious yawning and itching, and to squeamishness. The shared experiences may also extend to emotional states, and a 2018 study found some evidence that mirror-touch synesthetes are better able to read facial expressions and have higher levels of empathy than others.[33]

6 Agency

> She raised one hand and flexed its fingers and wondered how this thing, this fleshy spider on the end of her arm, came to be hers, entirely at her command. She bent her finger and straightened it. The mystery was in the moment before it moved, the dividing instant between not moving and moving, when her intention took effect. It was like a wave breaking. . . . There was no stitching, no seam, and yet she knew that behind the smooth continuous fabric was the real self—was it her soul?—which took the decision to begin movement and gave the final command.
>
> —Ian McEwan, *Atonement*

The second core component of bodily awareness, agency refers to the sense that we are in control of our body, our thoughts, and our actions. Agency and body ownership, the other core component of bodily awareness, are related but separate from each other and function largely independently. We know this because, even though these two senses can influence each other in various ways, they exhibit what neuropsychologists call "double dissociation"—that is, each one can be disturbed by itself while the other remains unaffected. Thus, in the condition commonly referred to as "alien hand syndrome," stroke patients with this condition acknowledge that their affected arm belongs to them, but believe that they are not in control of its actions, claiming instead that its movements are being controlled by some external force. That is, their sense of agency for the limb is disturbed, but their sense of ownership over it remains intact. On the other hand, stroke patients with somatoparaphrenia remain in full control of the affected limb, but strongly believe that it belongs to somebody else—body ownership is disrupted but agency is not. The sense of agency is closely tied to our notion of free will, and disturbances in how the brain processes agency-related information seem to explain various "paranormal" phenomena, such as how the ouija board works.

Invented in the late nineteenth century, the ouija board has the letters of the alphabet written on it, along with the numbers 0 to 9, and the words "Yes," "No," and "Good-bye" around the edges. It comes with a triangular pointing device called a "planchette" (or "little plank"), which is perched on the board. Two or more people sit around the board, place their fingertips on the planchette and ask questions, then follow the planchette as it moves around the board, apparently of its own volition, to spell out the answers. Ouija boards, typically presided over by a psychic during a séance, enable ghosts or spirits to communicate with the living from beyond the grave, or so those who trust in them believe. This belief is usually strengthened by the séance participants' experience that they are not moving the planchette themselves.

One explanation for how ouija boards work is through "facilitated communication," developed as a therapeutic technique in the early 1980s by a teacher named Rosemary Crossley as a way of communicating with people who have nonverbal autism, cerebral palsy, or other conditions that impair their ability to communicate. The technique involves having a trained "facilitator" sit with these individuals and hold their hand in order to support them in typing out words on a computer keyboard. In this way, Crossley and others reported that people who had never uttered a word were suddenly able to communicate meaningfully by typing out sentences and, sometimes, even full-length reports. This technique was discredited in the 1990s, however, by studies showing that the subjects' responses in fact originated with the facilitators themselves. In one 1993 study, researchers delivered different questions to the facilitators and subjects through headphones and received answers to questions asked of the facilitators; another study, in 1994, showed that subjects were unable to communicate facts that were unknown to the facilitator; and still another study, in 1995, showed that subjects were unable to describe objects they had seen previously in the absence of a facilitator.[1]

Just as séances reached the height of their popularity at the start of the twentieth century, one man and his horse were set to take the world by storm. In 1904, a German mathematics teacher named Wilhelm von Osten exhibited an Arab stallion from Russia named "Hans," who could, after four years of training, apparently perform all manner of intelligent feats. To the amazement of Berlin's general public, Hans could perform arithmetic and count how many members were in his audience, by tapping his hoof. He could also spell out the name of painters when shown works of art and the names of the composers of music he heard, using an alphabet in which the letters had been replaced by numbers, where A = 1, B = 2, and so on. Remarkably, Hans seemed able to do this even when questioned by people he had never seen before.

"Clever Hans," as he came to be known, garnered the attention of journalists and soon became a worldwide sensation. In August 1904, *New York Times* special correspondent Edward T. Heyn traveled to Berlin to see Clever Hans for himself, and he wrote a glowing full-page article about the performance. "For over four years Herr von Osten has given the animal systematic instruction such as he would a child," Heyn wrote. "Hans is an expert in numbers, even being able to figure fractions. He answers correctly the number of 4's in 8, in 16, in 30&c. When asked how many 3's there are in 7, he stamps down his foot twice [for the whole number] and for the fraction once. Then, when 5 and 9 are written under each other on the blackboard and he is asked to add the sum, he answers correctly." The article goes on to describe other astonishing abilities: "He can distinguish between straw and felt hats, between canes and umbrellas. He knows the different colors. One beholds several colored rags fastened on a string. A cavalry officer places himself before the horse and Hans is asked to state the color of his cap. The horse answers by stamping his foot down three times, the color of the third rag, which, like the cap, is red."[2]

Agency—a Sense of Control

Most of us experience the feeling of having voluntary control over our body, our thoughts, and our actions, the sense that we are in control of the movements of our body when we walk, talk, and perform all manner of other activities. Our thoughts and actions and the effects of our actions on the outside world appear to be seamlessly linked; we generate an intention to act, and we then move our body accordingly, to produce the desired effect on our surroundings. Thus, for example, I sit in an armchair reading a book; evening approaches and I sense the room getting darker, so I reach up to turn the lamp on, and when I flick the switch, light from the bulb floods the pages of my book.

The feeling that we are in control of our body, our thoughts, and our actions is referred to as the "sense of agency," which, along with the sense of body ownership, is the second core component of bodily awareness. The link between our thoughts, actions, and their consequences on the outside world is strong, but the underlying brain mechanisms are extremely complex. How do intentions form in the brain? How do we choose between different courses of action? And when do we become conscious of the choices? All of these questions remain unanswered.

Several studies indicate that the brain distorts our perception of time to enhance the feeling that we are in control of our actions. In a 2002 study, researchers at University College London (UCL) asked participants to judge when they perceived separate voluntary and involuntary actions and their effects to take place. In one experiment, participants pressed a key whenever they chose to, which produced a sound after a brief interval. The researchers then applied magnetic stimulation to the participants' motor cortex to induce muscle twitches that caused involuntary key presses, each of which again produced a sound. In both cases, participants were required to estimate when a key press and the resulting sound occurred, by noting the position of the second hand on a clock face.

In all cases, the interval between a key press and the resulting sound was exactly 250 milliseconds (or one-quarter of a second). But the researchers found that participants' perception of the interval shifted one way or the other depending on whether the key press was voluntary or involuntary. When they pressed the key of their own volition, they perceived their action to have begun fractions of a second later, and the resulting sound to have occurred fractions of a second earlier, than the action and sound actually did. Involuntary key presses elicited by magnetic stimulation had the opposite effect—they were perceived to have occurred slightly earlier, and the resulting sounds perceived to have occurred slightly later.

In other words, the brain had actively altered participants' perceptual awareness of the time at which these events took place. During voluntary action, the movement and its sensory consequences were perceived to have occurred more closely together in time, and this enhanced the participants' sense of agency over their actions—an effect the researchers dubbed "intentional binding." By contrast, the involuntary actions and their effects were perceived to have occurred farther apart, making the participants' more aware that they were not, in fact, in control of the actions. "These results suggest," the researchers conclude, "that the brain contains a specific cognitive module that binds intentional actions to their effects to construct a coherent conscious experience of our own agency."[3]

A more recent variation of this study, in 2015, showed that coercion alters the intentional binding effect. When coerced to perform the key presses, participants perceived the interval between key presses and their effects to be longer than when they performed them voluntarily. Thus coercion reduces our sense of agency, the feeling that we are in control of the actions we were coerced into performing—we were "only obeying orders."[4]

Researchers in Houston, Texas, provided an alternative explanation of how cause and effect are linked in the brain. Their 2006 study was designed in a similar way, but included a functional magnetic resonance imaging (fMRI) brain scanning experiment. Each time participants pressed a key, a flash of light appeared after a brief, fixed delay, which participants perceived to have been produced by their key press; subsequently, when flashes of light appeared at unexpectedly shorter delays *before* their key presses, these were perceived *not* to have been produced by their key presses. These findings suggest that sensations occurring at consistent delays after an action are perceived as being produced by that action, and that the brain not only recalibrates the timing of movements and sensations, but also generates a fixed representation of time against which the timing of current events is compared. The brain scanning experiment further revealed that this "reversal illusion" was associated with increased activity of the anterior cingulate cortex, a brain region known to play an important role in self-awareness and detecting conflicts of information.[5]

Alien Hands

Given the complexity of the brain mechanisms underlying the sense of agency, it will come as no surprise that a wide variety of conditions disrupt these mechanisms in one way or another, and there are numerous examples of this in popular culture. For example, the titular character in Stanley Kubrick's 1964 film *Dr. Strangelove* is an erratic, wheelchair-bound German who directs the US government's weapons research and development program. In the final scene of the film, while describing a postapocalyptic scenario to the president following a nuclear attack on Russia, his right hand takes on a life of its own—it raises itself into a Nazi salute, and then clutches at his throat, while he tries to restrain its movements with his other hand.

A similar but far more gruesome scene appears in another cult classic, Sam Raimi's *Evil Dead 2* (1988), in which the main character Ash's right hand is possessed by a demonic spirit that is tormenting him during a stay in a remote cabin in the woods. The possessed hand picks up plates and shatters them over his head, punches him in the stomach, grabs his hair, and smashes his head against the sink. When he loses consciousness and falls to the floor, it pulls him toward a nearby meat cleaver, but Ash wakes up just in time to pin the hand to the floor with a large knife, then sever it with a chainsaw.

Ridiculous as they may seem, these scenes have their basis in a mysterious neurological condition, usually called alien hand syndrome in the academic literature. First described in the late nineteenth century, this syndrome is very rare, occurring as a result of brain damage arising from a stroke or tumor. It is characterized by complex, goal-directed movements in one hand that are both involuntary and uncontrollable—typically, repetitive, unintentional grasping of objects that often involve "self-oppositional" behaviors, in which one hand counteracts the intended actions of the other.

In both early and more recent accounts of this intriguing syndrome, patients report that one of their hands seems to have a will of its own. One case report, published in 1905, describes a female German patient known as "H.M.," who claimed that "die Hand ist nicht normal; sie tut was sie selber will" (the hand is not normal; it does what it wants), and who believed that "das ein böser Geist in der Hand ist" (there is an evil spirit in the hand).[6]

A more recent case report, in 1994, describes the behavior of "Mrs. G.P." while out having dinner with her family and neurologist: "Out of the blue and much to her dismay, her left hand took some leftover fish-bones and put them into her mouth. A little later, while she was begging it not to embarrass her any more, her mischievous hand grabbed the ice-cream that her brother was licking. Her right hand immediately intervened to put things in place and as a result of the fighting the dessert dropped on the floor. She apologized profusely for this behavior that she attributed to her hand's disobedience. Indeed, she claimed that her hand had a mind of its own and often did whatever 'pleased it.'"[7]

A third case report, described in a 1991 longitudinal study, is that of a fifty-six-year-old woman referred to as "G.C.," who experienced a brain hemorrhage and fell into a twenty-day coma, following a stroke caused by a surgical procedure to "clip" two aneurysms in the anterior communicating artery of her brain. Several months after her operation, G.C. began to exhibit "severe disturbances of organization of voluntary gestures." For example, "the right hand frequently carried out complex activities that were not willed by G.C. . . . activities [that] were clearly goal-directed and . . . well executed, but undesired by the patient, who used her left hand to try to stop them . . . when the patient had a steaming cup of tea in front of her, the right hand proceeded to pick it up and bring it to her mouth, even though the patient knew that it was too hot and had just said she would wait a few moments until it had cooled. Nevertheless, it needed the intervention of her left hand to replace the cup on the table." Because she experienced these problems on a daily basis, G.C. came to consider "her left hand to be the one she could trust, while the right hand, which could make motions completely without her wanting it to, was the untrustworthy one that 'always does what it wants to do' . . . and in order to complete the activity she wanted to complete, she had to prevent the 'wayward' right hand from interfering, sometimes by force (by hitting it or by sitting on it)."

Occasionally, the "wayward" hand behaved in a socially or sexually inappropriate way. "On other occasions, when G.C. had a genital itch, the right hand scratched it vigorously, in view of other people, causing considerable embarrassment to the patient, who tried to stop the right hand with her left."[8] And a fourth case report, in a 2013 study, is that of a forty-five-year-old male patient who had had a stroke and who "complained of involuntary masturbation . . . [stating] that his right hand used to grab his genitals involuntarily, even in public places, often to his embarrassment and that of his wife";[9] and several other reports describe patients who involuntarily grabbed or fondled their genitals—and even those of other people.

Free Will

The sense of agency is closely tied to the notion of free will, our ability to act spontaneously and voluntarily. We have a strong feeling that we are free to choose between different courses of action, that we are agents of our own destiny, and this feeling is fundamental to our belief in our individuality. The notion that we are responsible for our actions lies at the heart of the criminal justice system; in law, a criminal action is defined both by the physical act of the crime (*actus reus*, meaning "criminal act"—literally, "wrongful deed") and by the conscious intent behind it (*mens rea*, meaning "criminal intent"—literally, "guilty mind"). People are punished for crimes they are judged to be responsible for, and when their responsibility is found to be diminished for some reason, they are judged to be not fully liable for their actions and punished less severely.

The view that our thoughts and feelings are nothing more than the product of electrical activity within a vast network of brain cells has become increasingly dominant. Yet our not knowing the steps between deciding upon a given course of action, forming the intention to act, and executing the act decided upon constitutes a huge gap in our understanding. For, even though philosophers have debated about free will for centuries, neuroscientists did not join the debate until the second half of the twentieth century, with a pair of seminal studies that produced surprising results.

Thus in 1965, neurophysiologist Hans Kornhuber of the University of Freiburg in Germany and his student Lüder Deecke used electroencephalography (EEG) to analyze the brain activity associated with self-initiated movements in humans. In a relatively simple experiment, they sat participants in a chair, attached electrodes to their scalp, over the motor and supplementary motor cortices of their brain, which are involved in executing and planning movements, respectively. They instructed their participants to flex one of the fingers on their right hand, whenever it suited them, up to 500 times at each sitting, consistently recording the electrical activity that *preceded* these voluntary movements. Kornhuber and Deecke named this electrical activity the "readiness potential" (*Bereitschaftspotential*), which they defined as "the electrophysiological sign of planning, preparation and initiation of volitional acts."[10]

Some twenty years later, in 1983, Benjamin Libet and his colleagues repeated this experiment at the University of California, San Francisco, and obtained very similar results. Again, the researchers instructed their participants to flex the fingers or wrist of their right hand, whenever they felt like doing so. In line with Kornhuber and Deecke's 1965 results, they observed electrical activity in the brain that preceded the participants' voluntary movements. This experiment went one step further, though: Libet and colleagues asked the participants to report when they had decided to perform each movement by noting the position of a hand moving around a clock face. The researchers' results showed that the readiness potential preceded not only the participants' movements, but also their reports of when they had formed the conscious intention to perform them, by an average of 350 milliseconds. This led Libet and his colleagues to conclude that "cerebral initiation of a spontaneous, freely voluntary act can begin unconsciously, that is, before there is any (at least recallable) subjective awareness that a 'decision' to act has already been initiated cerebrally."[11]

In other words, actions that we think of as being voluntary are in fact initiated unconsciously. These results caused quite a stir and are still interpreted by many as meaning that free will is nothing more than an illusion. Libet and colleagues' experiment has been repeated, and its results replicated, using various other methods. In a 2008 study, for example, one research team used fMRI to detect supplementary motor and prefrontal cortex activity associated with the intention to press one of two buttons a full 10 seconds before participants reported making the decision of which button to press.[12] A study published three years later, in 2011, involving microelectrode recordings from more than 1,000 neurons in the brains of twelve conscious epilepsy patients undergoing surgical evaluation, showed that individual cells in these same brain regions fired up to 1.5 seconds before the patients reported making the decision to press a key on a laptop with their index finger. What's more, from the activity of these brain cells, the researchers could predict not only when their patients would perform a key press to within a few hundred milliseconds, but also whether the patients would use their left or right index finger to press the key.[13]

Another line of evidence, as reported in a 2012 study, suggests that Libet and colleagues' interpretation of their findings was incorrect, and that the readiness potential does not cause our actions at all. The brain's electrical activity is very "noisy," being filled with the waxing and waning of spontaneous fluctuations produced by random firing of large, discrete neuronal populations. When selecting between different courses of action, the brain weighs up the options available to it, with different groups of neurons gathering information about each possible course of action. When one of these groups accumulates enough evidence, its activity level passes a threshold, and the commands for the movements associated with that particular course of action are issued. Thus, rather than being a neural sign for the unconscious initiation of voluntary movement, the readiness potential may instead reflect the brain's commitment to one particular course of action from among all those available to it at that given moment in time.[14]

Action Models

Planning and performing movements are fundamental functions of the brain to which large portions of the cerebral cortex are devoted. The ability to control our movements accurately is essential for many aspects of daily life—from navigating complex environments, to reaching for and grasping objects, to learning new skills. It is an ability that we usually take for granted, perhaps because we perform so many of our movements effortlessly, without thinking about them, or at least without *consciously* thinking about them. Only when the underlying brain mechanisms break down do we realize how complex the processes needed to perform our movements are.

Neurologists have theorized about voluntary action for at least a hundred years, in part to explain a group of motor disorders collectively referred to as "apraxia," which usually occur as a result of brain injury or disease. Patients with apraxia are unable to carry out simple movements; in the clinic, they cannot handle objects in the right way, fail to produce the correct movements in response to spoken instructions, and cannot imitate movements they observe. These symptoms are sometimes accompanied by speech defects, as described by John Hughlings Jackson in 1861:

> In some cases of defect of speech the patient seems to have lost much of his power to do anything he is told to do, even with those muscles that are not paralysed. Thus, a patient will be unable to put out his tongue when we ask him, although he will use it well in semi-involuntary actions—for example, eating and swallowing. He will not make the particular grimace he is told to do, even when we make one for him to imitate. There is power in his muscles and in the centers for coordination of muscular groups, but he—the whole man, or the "will"—cannot set them a-going. . . . In a few cases patients do not do things so simple as moving the hand (i.e., the non-paralysed hand) when they are told . . . A speechless patient who cannot put out his tongue when told will sometimes actually put his fingers in his mouth as if to help get it out; and yet, not infrequently, when we are tired of urging him, he will lick his lips with it.[15]

To the German neurologist and psychiatrist Hugo Karl Liepmann, apraxia represented an inability to plan movements. Liepmann's first paper on the subject, in 1900, described a forty-eight-year-old senior civil servant who continued to experience difficulties moving his right hand long after he had had a stroke. He could fasten a button when his fingers were placed on it, but he could not move on to the next button and repeat the action. He also performed or copied simple gestures clumsily, and he failed to recognize everyday objects, although his sense of sight was largely intact. Liepmann also treated other patients with damage to the left hemisphere of their brain and found that they had apraxia in the unparalyzed right hand. But, in contrast, patients with a damaged right hemisphere did not have left-hand apraxia. Liepmann therefore concluded that apraxia occurs as a result of damage to the left hemisphere—which was later confirmed in autopsies— but that the left hemisphere must contain memories needed for organizing movement of the left as well as right hand. In order to execute a skilled movement, Liepmann theorized, the movement plan must first be retrieved and then transmitted to other areas of the brain.[16]

Paul Schilder recognized the importance of this idea. "The greatest progress made so far in the understanding of a human action is due to Liepmann's investigations," Schilder wrote in *The Image and Appearance of the Human Body*. "He has shown that every action is based on an anticipatory plan . . . [with] a specific structure. It not only contains the final aim, but also comprises the insight into the single actions which are necessary for the actualization of the plan." Schilder also recognized that "knowledge of one's own body is an absolute necessity," in this process, but he argued that "it would be wrong to believe that this plan exists in the full light of consciousness."

Our movements are controlled by a vast distributed network spanning at least three lobes of the brain. Neuroscientists refer to this network as the "motor control system," and some theorize about how models within this system represent our movements. Taking, as an example, the action of our picking up a glass, raising it to our mouth to drink from it, and then replacing the glass on the tabletop, numerous factors have to be taken into account. The control system must first determine the actual state of our body, or the current position of our limbs in space, as well as the desired state (or states) at each step of the process. It must then select the appropriate set of muscle commands, from a huge repository of possible commands, and then execute these commands. While performing the movements, the system monitors sensory feedback, both from within our body and from the outside world, to ensure that our movements are accurate, in order to achieve the desired outcome.

It is now widely believed that the brain makes predictions about the outcomes of actions, or the sensory consequences of each movement we make, based on representations of certain aspects of the body and the external world. Every time the system issues a command to execute a movement, it simultaneously produces a "forward model" that predicts the outcomes of that movement, and then compares its predictions to the actual outcome of the movement. In this way, the motor control system can anticipate and compensate for the changes of sensation produced by the movement, filter out the sensations produced both by our own movements and by changes in the outside world, and maintain accuracy of movement when sensory feedback is delayed or absent. According to this "comparator model," comparison between the predicted and desired state gives us a feeling of being in control of our actions, and comparison between the predicted and actual outcome allows us to attribute the outcome to ourselves.[17]

Thus the control system likely contains multiple motor representations, which interact with one another when we plan and perform movements. Only some of these representations are available to our conscious awareness, however. We are conscious of initiating and controlling our movements, but some aspects of these processes take place without our awareness. We can imagine performing movements in our "mind's eye," and neuroimaging studies show that imagining movements activates some of the brain areas involved in executing those movements. Professional athletes use mind's eye visualization techniques to rehearse their movements; this motor imagery can help them improve their performance, even without physical training. But an estimated 1 to 3 percent of people apparently lack the mind's eye altogether—they cannot visualize or conjure up mental images, a condition now known as aphantasia.[18]

Neurosurgical patients and those with brain damage provide more clues about voluntary action and motor imagery. In patients undergoing awake brain surgery, electrically stimulating the right inferior parietal cortex (which lies just behind the top of the ear) triggered a strong desire in these patients to move their left hand, arm, or foot, eliciting spontaneous reports that included words such as "will," "desire," and "wanting to" move, whereas electrically stimulating the same area in the left inferior parietal cortex evoked a strong intention in patients to move their tongue or lips or to talk, and when this area in the left hemisphere was stimulated more intensely, the patients believed that they actually had performed these movements or actions even when they had not ("I moved my mouth, I talked, what did I say?"). And stimulating the supplementary motor cortex caused movements of both mouth and limbs, which the patients then denied. Damage to the right parietal lobe also seems to impair motor imagery of both hands.

Together, these findings suggest that the supplementary motor area of the brain plays a role in generating the intention to move, whereas the parietal cortex brings the intention into conscious awareness and also plays an important role in generating the representations involved in motor imagery.[19]

Motor Control Malfunctions

Malfunctions in the various components of the motor control system give rise to all manner of unusual symptoms and behaviors that can be explained, at least to some extent, by what we have come to know about the system's components and their workings.

An important function of our sense of agency is to help us distinguish between our own actions and those of others, a distinction that can break down in schizophrenia. People diagnosed with schizophrenia very often describe the feeling that their actions, thoughts, and speech are being controlled by some external force, rather than by themselves. Auditory and visual hallucinations are common symptoms of schizophrenia and can be thought of as perceptual errors in which internally generated sights and sounds are believed to arise in the outside world. Schizophrenia patients may also believe that their behaviors do not match their intentions, saying things like "My fingers pick up the pen, but I don't control them; what they do has nothing to do with me," for example, or "The force moved my lips. I began to speak. The words were made for me."[20]

Such delusions of alien control could arise for a number of reasons, with most explanations revolving around motor representation deficits that interfere with internal self-monitoring processes. According to one hypothesis, patients misattribute agency to external forces because of abnormalities in their motor imagery. Many human movement studies involve recording the speed and accuracy of a specific hand movement, such as pointing at a target area of a visual stimulus. Typically, there is a trade-off between speed and accuracy, such that the smaller a target area is, the longer the pointing movement takes, and this holds true for imaginary as well as actual movements. For schizophrenia patients who do not experience delusions of control, target size affects their performance of imaginary movements just as it does that of nonschizophrenia study controls, but for those schizophrenia patients who do experience delusions of control target size does *not* affect the time it takes them to make the imaginary movements. Thus delusions of control in schizophrenia appear to be related to an impaired ability to distinguish changes in a visual stimulus while executing an imaginary movement.[21]

Other research suggests that schizophrenia patients make inaccurate predictions about the outcomes of their actions. In a series of experiments in 2010, a group of twenty schizophrenia patients and twenty nonschizophrenia matched controls were asked to make pointing movements in a virtual reality setting where a computerized image of their finger was displayed on a monitor via a mirror. The apparatus was set up such that the participants saw the virtual image at the same level as their actual finger. Using their right index finger to point freely at any position within the upper right quadrant of a circle, the participants received visual feedback from a cursor whose movements corresponded to the tip of their finger. The researchers manipulated this feedback during some of the pointing movements by rotating the cursor to varying degrees in one direction or another with respect to position of the participants' fingertip, thus they could vary the accuracy of the visual feedback and determine exactly when each participant realized that the feedback was being distorted.

The researchers found that the schizophrenia patients had a significantly higher threshold for detecting the experimentally induced feedback distortions than the control group had. Whereas the controls noticed small discrepancies between their finger movements and those of the feedback cursor, the schizophrenia patients noticed only discrepancies that were significantly bigger. What's more, the patients' detection threshold was related to the intensity of their delusions—the stronger their delusions, the greater was their impairment in detecting the feedback distortion, and the bigger were the discrepancies before they noticed them. The researchers interpreted this to mean that schizophrenia patients' inaccurate predictions of the sensory consequences of their actions are the source of their delusions of control, causing them to more readily misattribute their own actions to external forces.[22]

Various other studies show, however, that schizophrenia patients' sense of agency is not necessarily dependent upon their ability to predict action outcomes, and that agency may be felt even during the process of selecting actions, rather than after an action has been performed. By contrast, in another 2010 study involving nonschizophrenia participants, researchers used subliminal cues to influence the participants' decision to press one of two keys, and found that this "priming" did affect the participants' sense of control over the consequences of the key presses. It made them more likely to perform the actions primed by the preceding subliminal cues, and they felt more control over the effects of key presses that were compatible with the cues than over those which conflicted with them.[23]

These and other findings have led some to reconsider the comparator model described above and to develop a more complex model, one that explains the sense of agency as a two-step process involving multiple factors. According to this model, the brain mechanisms underlying our sense of agency are divided into two separate processes: (1) a low-level, unconscious feeling of agency, involving the integration of different types of information, such as the forward model of action outcomes, and different types of sensory feedback, and (2) a high-level judgment of agency, which enters our awareness and allows us to interpret the low-level information on the basis of our intentions and various cues from the outside world. Errors in one of the two steps could lead to an overreliance on information from the other step and thus to misattributions of agency.[24]

Malfunctions of the motor control system also help to explain the alien hand syndrome. Autopsies and brain scans have revealed that the syndrome is often associated with damage to the corpus callosum, the massive bundle of approximately 100 million nerve fibers that connects the left and right hemispheres of the brain and that allows them to pass information back and forth. To understand how a damaged corpus callosum might give rise to alien hand syndrome, we first have to understand two basic principles of brain structure and function: contralateral control and lateralization of cerebral function.

The brain and nervous system of humans and many other animals display "contralateral control," with the left hemisphere of the brain controlling the right side of the body, and the right hemisphere the left side. Axonal fibers of neurons in the primary motor cortex bundle together to form a pathway called the "corticospinal tract," which descends in the spinal cord to connect with "secondary motor neurons"; these, in turn have axons that project and send signals to the muscles of the body. The vast majority of the axonal fibers in the primary motor cortex cross over from one hemisphere of the brain to the other, even before entering the spinal cord, in the medulla oblongata, at the bottom of the brain stem. We do not know why the nervous stem is organized in this way; what is clear, though, is that this feature has been found to be present in all vertebrate animals—mammals, birds, reptiles, amphibians, and fish—and must have arisen hundreds of millions of years ago in some common evolutionary ancestor.

The lateralization of cerebral function refers to the tendency for certain of the brain's functions to be performed in specialized regions found in one hemisphere or the other, language being the best-known example. In right-handed people, who make up about 90 percent of the world's population, the language centers are located in the left hemisphere (although we now know that areas in the right hemisphere still perform various language-related functions); left-handers, by contrast, appear to have language centers represented more equally in both hemispheres of the brain.

Alien hand–type behaviors most often arise on the left side of the body for unknown reasons. This led some early observers to explain the alien hand syndrome in terms of a disconnection between the right motor cortex, which commands the left hand, and the left motor cortex. Conscious language processing is typically localized in the left hemisphere, so disconnecting the two hemispheres of the brain would lead to a situation where intentions and control were processed separately from each other. If this explanation is correct, then right-handed people should only ever experience left-handed alien behaviors. Because, however, neurologists have identified several right-handed patients with right-sided alien hands, this "interhemispheric disconnection" explanation would seem to be inadequate.

Closer examination reveals that damage to the front of the corpus callosum often encroaches upon adjacent areas of the frontal lobe on the side of the brain opposite that of the alien hand, particularly the supplementary motor cortex, which is known to be involved in movement planning and execution and thought to be involved in translating our intentions into self-initiated voluntary actions. And because the supplementary motor cortex is activated not only during movement preparation, but also when movements are imagined, and activity in some of its subregions appears to drop sharply just before movements are executed, it may play a role in inhibiting the execution of certain movements. Damage to this area of the cortex would therefore prevent this inhibitory function, unleashing the alien hand movements.[25]

Alien hand syndrome is also associated with damage to the back of the corpus callosum, and patients having this kind of damage present with quite different symptoms from those having damage nearer the front of the structure. Patients with damage to the front of the corpus callosum experience complex intentioned movements that seem to be performed against their will, whereas those with damage to the back of the corpus callosum experience something quite different. "Alien hand syndrome" is therefore a misnomer, used incorrectly to describe several different conditions. The term dates back to a 1972 paper published in French in the *Revue Neurologique*, describing four patients who presented with the same set of behaviors due to tumors at the back of the corpus callosum: "The patient who holds his hands one within the other behind his back does not recognize the left hand as his own. . . . The sign does not consist in the lack of tactile recognition of the hand as such, but in the lack of recognition of the hand as one's own." The study authors, Serge Brion and Charles-Pierre Jedynak, referred to this phenomenon as "main étrangère," which they translated as "strange hand" in the English abstract of their paper. Subsequently, however, others writing about the behavioral consequences of damage to the corpus callosum translated "main étrangère" as "alien hand," and the name stuck.

"Alien hand," then, refers to the syndrome caused by damage to the back of the corpus callosum, which results in partial hemisomatognosia, or the inability to recognize a body part on the side opposite that of the damage. This occurs because the brain damage extends into the parietal lobe, which plays an important role in bodily awareness and likely contains "higher-order" body representations generated by multisensory integration. Alien hand syndrome is thus a disorder of body ownership, not one of agency, and some neuropsychologists propose that the disorder popularly called "alien hand" or "Dr. Strangelove syndrome" should be renamed the "anarchic hand sign."[26]

The Ideomotor Effect

But what of Clever Hans and ouija boards? In 1904, the German Board of Education set up a special commission to investigate the claims about Hans, and found no evidence of fraud. "The remarkable horse called "Clever Hans" has just been examined by a special commission of experts, in order that a decision might be arrived at whether it is a horse possessed of extraordinary brain power or merely, like many others of its tribe, peculiarly adapted to learning tricks from patient trainers," reads a short news story published in *The London Standard,* and syndicated in the *New York Times* on October 2. "The commission consisted of the well-known circus proprietor, Herr Paul Busch; Count Otto zu Castell Ruedenhausen, a retired army Captain; Dr. Grabow, a retired schoolmaster; Dr. Ludwig Heck, Director of the Berlin Zoological Gardens; Major von Keller, Major Gen. Koering, Dr. Miessner, a veterinary surgeon; Prof. Nagel of the Physiological Institute of the University of Berlin, and several other prominent men. . . . [It] has issued a statement declaring that it is of the opinion that there is no trickery whatever in the performances of the horse, and that the methods employed by the owner . . . differ essentially from those used by trainers, and correspond with those used in teaching children in elementary schools."[27]

Not all experts were convinced, however. In 1907, a comparative biologist and psychologist named Oscar Pfungst reported that Hans could not answer questions that his trainer or other questioner did not know the answer to, and could not answer anything when a screen was placed between him and his examiner. Eventually, it became clear that Hans had learned to read microscopic cues in his master's face, which indicated that he had tapped his hoof the correct number of times. These signals are transmitted unconsciously—even Pfungst himself was unable to control them, and inadvertently gave Hans the answers to his questions. Willhelm von Osten, the teacher who trained Hans, died in 1909, but the fate of Hans is unclear. He was drafted as a military horse when World War I broke out in 1914 and reportedly killed in action two years later.[28]

There are at least three types of movement that are performed outside our conscious awareness. One such type, excitomotor actions, which control breathing and swallowing and which keep our hearts beating, are almost always performed unconsciously, although some of them can enter our awareness under certain circumstances. Sensorimotor actions are another type—the knee-jerk reflex, for example, or the withdrawal reflex, which occurs when we touch a painfully hot surface. The third type, ideomotor actions, are voluntary movements we perform without being consciously aware of them.

Ideomotor actions occur when the neural circuits that plan, select, and execute motor commands function independently of those which bring our actions into conscious awareness. Under these circumstances, our voluntary actions are guided not by sensations from the body and outside world, but by ideas and expectations (hence the prefix "ideo-"). Because the mechanism of action selection and execution is disengaged from the mechanism of conscious action control, we have no sense of agency over our actions, and thus deny any responsibility for them.

Ideomotor actions were first described by the physician and physiologist William B. Carpenter in a lecture at the Royal Institution in London in 1852, delivered with the intention of providing a scientific explanation for how ouija boards produce their effects and also for how the unconscious signals were conveyed to Clever Hans:

> Now the usual *modus operandi* of sensations is to call forth *ideas* in the mind; and these ideas . . . become the subjects of intellectual processes, which result at last in a determination of the will. The movements which we term *voluntary* or *volitional* differ from the emotional and automatic, in being guided by a distinct conception of the object to be attained, and by a rational choice of the means employed. And as long as the Voluntary power asserts its due predominance, so long can it keep in check all tendency to any other kind of action, save such as ministers directly to the bodily wants, as the automatic movements of breathing and swallowing. . . .
>
> . . . when *ideas* do not go on to be developed into emotions, or to excite intellectual operations, they, too, may act (so to speak) in the transverse direction, and may produce respondent movements. . . . Here the movements express the idea that may possess the mind at the time; with these ideas, emotional states may be mixed up, and even intellectual operations may be (as it were) automatically performed under their suggestive influence. But as long as these processes are carried on without the control and direction of the Will, and the course of thought is entirely determined by suggestions from without, (the effects of which, however, are diversified by the mental constitution and habits of thought of the individual) such movements are as truly automatic, as are those more directly prompted by sensations and impressions, although originating in a more truly psychical source. But the automatic nature of the purely emotional actions can scarcely be denied; and it is in those individuals in whom the intellectual powers are the least

> exercised, and the controlling power of the Will is the weakest, that the emotions exert the strongest influence on the bodily frame, so we may expect Ideas to act most powerfully when the dominance of the Will is for the time suspended.[29]

Action-Perception

"Perceiving is a way of acting," writes the philosopher Alva Noë. "Perception is not something that happens to us, or in us. It is something we do. . . . The world makes itself available to the perceiver through physical movement and interaction. . . . Perceptual experience acquires content thanks to our possession of bodily skills. *What we perceive* is determined by *what we do*."[30]

The brain is a "black box" whose workings are deeply mysterious to us. We used to view it in "behaviorist" terms as an input-output system, which gathers information about the world through the sense organs (the input), and then processes it to generate a response (the output), in the form of movements. According to this view, information processing occurs in sequential stages, beginning with perception and ending with action: we see, we think, and then, finally, we act. The reality is far more complex, however. Action and perception are inextricably linked, and can be regarded as opposite sides of the same coin. We perceive to act, and we act to perceive.

Nearly one hundred years ago, Wilder Penfield observed that not only do the primary sensory and motor cortices of the brain lie next to each other, but they also work in unison. Electrical stimulation of certain subfields of the somatosensory cortex evokes movements in the corresponding body parts, whereas stimulation of the motor cortex evokes touch sensations. The somatosensory cortex is mostly concerned with perception, and the motor cortex with action, yet they do not work independently, but as a single functional unit. Action and perception influence each other in ways that we are only just beginning to understand. Although we act in response to the external stimuli we perceive, and everything we do alters the way we perceive our body and the outside world, the relationship between action and perception works both ways.

In a classic 1963 experiment, psychologists Richard Held and Alan Hein raised ten pairs of kittens in the dark and regularly placed each pair in a carousel apparatus for brief periods of time. The apparatus consisted of a circular drum decorated with vertical stripes, and furnished so that one kitten was free to walk around it while the other sat in a small metal basket. The kittens were coupled so that when one walked around the carousel, the one in the basket moved with it. Thus all the kittens were exposed to exactly the same kind and amount of visual stimulation, but for those in the basket, perception and action were not directly linked. Afterward, when Held and Hein tested the kittens, they found that the depth perception of those who had been placed in the basket was severely impaired. When placed onto a surface, the "active" kittens extended their paws in anticipation, but the "passive" ones did not. And when placed in the middle of a "visual cliff" consisting of a sheet of glass over patterned surfaces of differing heights, the active kittens always walked off the "shallow" end, whereas the passive ones walked off the deep and shallow ends at random, which again suggests that they were unable to perceive depth cues. From this, Held and Hein concluded that "self-produced movement with its concurrent visual feedback is necessary for the development of visually guided behavior."[31]

Perception is itself a form of action. That is, perception is an active not a passive process; rather than merely receiving sensory inputs from

Figure 6.1
The Rubin Vase illusion is an "ambiguous" figure that can be perceived as either a vase or two faces in silhouette.

the outside world, in perceiving, the brain *acts upon* this information as it receives it. Visual illusions demonstrate this clearly. Some illusions are ambiguous figures, which can be interpreted in one way or another, but never in both ways at the same time. One famous example is the Rubin vase, which can be perceived either as a vase or as two opposing faces in silhouette. When we look at such an illusion, we see what we expect to see. Our prior knowledge, expectations, and biases are brought to bear upon the most basic processes of sensation and perception.

Our actions influence the way we perceive the world, and the way we perceive the world influences both the way we think about the world and our ability to act within it. When athletes and sportspersons are performing well, they seem to perform effortlessly, but when they are performing badly, their actions seem to be so much harder. A soccer player who has just scored perceives the goal to be bigger than it actually is, making it easier for the player to score again, but a player who repeatedly misses the goal after scoring once perceives it to be smaller and finds it harder to score again. Thus the greater the *perceived* effort required to achieve any goal, the harder it will be to achieve. This general rule applies beyond the playing field: a hill seems steeper when we are tired, or hungry, or when we are carrying a heavy backpack, and a journey to an unfamiliar destination seems so much longer.[32]

7 Distortions

Bodily awareness is based on the brain's representations of the body, which in turn are largely based on a continuous influx of sensory information entering the brain from the body. Consequently, interruptions in the stream of sensory information or disturbances in how the brain processes that information can alter perception of the body. Anorexia is one relatively common example of this. Growing evidence suggests that this condition involves distorted body representations, which alter how those with anorexia perceive and, perhaps more importantly, think about their body. Pain and anesthesia can also alter how we perceive our body, as can various narcotic drugs. Descriptions of body image distortions are also found in the classics of literary fiction, most notably in *Alice's Adventures in Wonderland.*

In Lewis Carroll's book *Alice's Adventures in Wonderland,* Alice experiences a series of sensations in which her body seems to change in size and shape. Alice's adventures start when she falls down a rabbit hole; finding herself in a long, dark hall with doors on every side, she manages to unlock one of them and sees that it leads to a small passage, "not much larger than a rat-hole." Kneeling down, she looks along the passage into a lovely garden. Alice then finds a little bottle labeled "DRINK ME," and when she drinks its contents, her body begins to shrink: "'What a curious feeling!' Said Alice. 'I must be shutting up like a telescope!' And so it was indeed: she was now only ten inches high, and her face brightened up at the thought that she was now the right size for going through the little door into that lovely garden."

Next, the shrunken Alice finds a little box containing a very small cake, on which currants mark out the words "EAT ME." When she does so, her body grows to "more than nine feet high," and her head strikes the ceiling: "'Curiouser and curiouser!' cried Alice, 'now I'm opening out like the largest telescope that ever was! Good-bye, feet!' [figure 7.1] (for when she looked down at her feet, they seemed to be almost out of sight, they were getting so far off). 'Oh, my poor little feet, I wonder who will put on your shoes and stockings for you now, dears?'"

Figure 7.1
In *Alice's Adventures in Wonderland,* Alice experiences a series of body image distortion hallucinations: "'now I'm opening out like the largest telescope that ever was! Good-bye, feet!'" Illustration by Sir John Tenniel.

Later on, after shrinking again, Alice encounters a large caterpillar sitting on top of a mushroom, "with its arms folded, quietly smoking a long hookah," who tells her that eating from one side of the mushroom will make her grow taller, and eating from the other will make her grow shorter. Not knowing which side is which, Alice stretches her arms around the mushroom, and breaks off a piece with each hand. When she nibbles one piece, her torso shrinks, and she feels her chin striking her foot. Taking a bite out of the other causes her neck to elongate to such a length that a pigeon flying overhead mistakes her for a worm.[1]

Alice in Wonderland Syndrome

Some forty years after *Alice's Adventures in Wonderland* was first published in 1866, accounts of hallucinations similar to those described by Carroll began to appear in the medical literature. In 1904, William Spratling, one of the first American epileptologists, published case studies of several patients for whom "everything looked bigger" just before their seizures; three years later, in 1907, the great British neurologist William Gowers also reported epilepsy patients who perceived objects to look "twice their size" during the aura preceding their seizures; and in 1913, the German neurologist Hermann Oppenheim noted that he had "seen a case of genuine hemicrania ["one-sided headache"] in which there was during an episode of violent migraine an indescribable feeling of detachment of the trunk or extremity after an hour or even a day of spontaneous dizziness."[2]

The American neurologist Caro Lippman noted, in a 1952 paper published in the *Journal of Nervous and Mental Disease,* that "the great variety" of hallucinations experienced during the migraine aura were still "little known to the medical profession." He incorrectly claimed that "there is no description in the migraine literature of hallucinations of the sense of body image," adding that "over a period of eighteen years of intensive migraine studies I have collected many histories of such hallucinations from both men and women," and he then went on to describe seven cases.

One case is that of a thirty-eight-year-old housewife whose headaches began during, and had recurred ever since, her second pregnancy, at the age of nineteen. "Some hours before the attack of one-sided headache and vomiting, and often during and after attack she may teeter or reel as though drunk," Lippman reports. "With these symptoms often occurs a sensation of the neck extending out on one side for a foot or more; at other times her hip or flank balloons out before, with, or after the headache. Very occasionally she has an attack where she feels small—'about one foot high,' [but] she says she knows the distortion isn't real because she looks in the mirror to see."

Another is that of a woman in her nineties who "stated that she had had classic migraine headache with nausea and vomiting . . . from childhood," who "complains frequently of her left ear "ballooning out 6 inches or more" a few hours before onset of a mild migraine headache. This feeling of ear distortion did not bother the patient, however, because she, too, "could see in the mirror that it did not exist."

In a third case, a twenty-three-year-old secretary described her hallucinations in a letter to Lippman: "About every six months I would have a major attack that lasted for weeks and required hospitalization. It was at these times that I experienced the sensation that my head has grown to tremendous proportions and was so light that it floated up to the ceiling, although I was sure it was still attached to my neck. . . . This sensation would pass with the migraine but would leave me with a feeling that I was very tall. When walking down the street I would think I would be able to look down on the tops of others' heads, and it was very frightening and annoying not to *see* as I was *feeling*. The sensation was so real that when I would see myself in a window or full-length mirror, it was quite a shock to realize that I was still my normal height of under five feet."

Another case is that of a thirty-eight-year-old woman whose hallucinations gave her "a very peculiar feeling of being very close to the ground as I walk along . . . as though I were short and wide, as the reflection in one of those broadening mirrors one sees in carnivals." Lippman adds that "if this attack occurs while [this patient] is returning from the grocery store which is at the bottom of the hill on which she lives, the top of the hill seems 'very far away.'"

Another patient described to Lippman "this same feeling of being short and wide" with explicit reference to Lewis Carroll, calling it "her 'Tweedle-Dum or Tweedle-Dee feeling,'" because it reminded her of the barrel-shaped twin creatures in *Through the Looking Glass and What Alice Found There*, the sequel to *Alice's Adventures in Wonderland*. The remaining case reports likewise include descriptions that sound remarkably like those in Carroll's *Wonderland* book: "the illusion of being taller than I actually am in relation to ordinary objects. . . . My head would seem far above my hands [or] much larger than the rest of my body"; "my neck stretches and my head goes to the ceiling"; "my body is as if someone had drawn a vertical line separating the two halves. The right half seems to be twice the size of the left half. I wonder how I am going to get my hat on when one side of my head is so much bigger than the other. After a few minutes of feeling large, the right half seems to shrink until it is smaller than the left."

Patients who reported such hallucinations had hitherto often been dismissed as delusional, but Lippman astutely noted the similarity of their experiences to Alice's as described by Carroll: "I would hesitate to report these hallucinations which I have recorded in my notes on migraine had not, more than 80 years ago a great and famous writer set them down in immortal fiction form," he wrote in concluding his 1952 paper. "*Alice in Wonderland* contains a record of these and many other migraine hallucinations. Lewis Carroll . . . was himself a sufferer from classic migraine headaches."[3]

Subsequently, in 1955, the English psychiatrist John Todd notes the similarity of "bizarre disturbances of the body image" in the hallucinations of epilepsy patients and patients having migraine headaches and "proposes to describe the experiences of these patients under the general heading 'the syndrome of Alice in Wonderland.'"[4] Some have suggested that Carroll experienced such body image distortions himself, and that they inspired *Alice's Adventures in Wonderland,* but this claim was challenged by an examination of his diaries, which found no entries referring to migraine until twenty years after he wrote the *Alice* books. A drawing and diary entry predating the books have since been discovered, however; both describe migraine symptoms, albeit not the ones described so vividly in *Alice's Adventures in Wonderland.*[5]

Despite this long history, Alice in Wonderland syndrome remained obscure until relatively recently. In the past two decades or so, scientists and clinicians have started to pay more attention to it, due partly to advances in functional neuroimaging technology, which enable them to investigate the relationship between symptoms and brain activity. The early reports typically described the syndrome symptoms as "hallucinations," but today they are more accurately described as "distortions of visual perception and body representations" arising from a "perceptual disorder." Another defining characteristic of this disorder is a distorted perception of time, which

Carroll also described: “The rabbit-hole went straight on like a tunnel for some way, then dipped suddenly down . . . [and Alice] found herself falling down what seemed to be a very deep well. Either the well was very deep, or she fell very slowly, for she had plenty of time as she went down to look about her, and to wonder what was going to happen next.”

Alice in Wonderland syndrome is thought to be very rare—fewer than 200 case descriptions have been published in the medical literature since Todd named it as such in 1955. The vast majority of these cases involve children, the average age of them being nine. In children, the syndrome is most often associated with encephalitis caused by infection with the Epstein-Barr virus; in adults, migraine is the most common cause, with the syndrome occurring in approximately 15 percent of those with migraines. Other causes include brain tumor, brain hemorrhage, scarlet fever, stroke, depression, and schizophrenia; and in 2011, doctors in Israel reported the case of an eleven-year-old who developed Alice in Wonderland syndrome after being infected with swine flu (“H1N1 influenza”). The syndrome has also been reported during sensory deprivation, as well as during hypnotherapy and the altered states of consciousness that occur just before falling asleep and just before waking (the “hypnagogic” and “hypnopompic” states).

Lewis Carroll accurately depicted some of the most common symptoms of the syndrome, namely, the feeling that one’s body is larger or smaller than it actually is (“macro-” or “microsomatognosia”) and objects appearing larger or smaller than they actually are (“macro- or micropsia”). But patients have reported myriad other symptoms, including the inability to perceive color or motion, enhanced depth perception, illusory movement, the illusion that objects have been split vertically, objects appearing flattened and elongated, objects appearing rotated by 90 or 180 degrees, and seeing multiple images as if looking through an insect’s compound eye. Usually, such symptoms are not long lasting, disappearing within a few minutes or days, either spontaneously or after treatment of the underlying cause; in cases of migraine and epilepsy, however, they may persist for years, or even throughout the patient’s lifetime. One or more of these individual symptoms are experienced more commonly in the general population, with one 1999 study showing that over one-third of the 297 adults sampled had experienced two such symptoms over the course of their lifetime.[6]

Anorexia

The distortions of body image experienced in Alice in Wonderland syndrome are usually consequences of some other affliction and can be disorienting, or perhaps a little frightening, but are otherwise harmless. They can, however, be a root cause, rather than a consequence, of other conditions, and, in some cases, they may be damaging, or even life threatening.

A prime example is anorexia nervosa, first described in 1689 by the English physician Richard Morton, who after examining an extremely emaciated eighteen-year-old female patient, described her as "a skeleton only clad in skin." And because "the wasting of the body" was "attended with a want of appetite," Morton called the condition "nervous consumption."[7] The term "anorexia nervosa"—which translates as "nervous lack of appetite"—was coined almost two centuries later, by the prominent English physician William Gull, who published a study in 1873 that detailed the symptoms, appearance, and behavior of three patients under his treatment who had this condition. Gull believed that anorexia affected both sexes, and he listed its characteristic symptoms as fatigue, general weakness, and loss of body mass; but he also pointed out that the condition mostly affected young women of reproductive age, and so included "amenorrhea," or lack of a menstrual period, as one of its chief characteristics.[8]

Almost at the very same time in 1873, the French neuropsychiatrist Ernest Charles Lasègue published his own description of anorexia. "The hysterical subject, after some indecision of but short duration, does not hesitate to affirm that her only chance of relief lies in an abstinence from food; and, in fact, the remedies appropriate to other gastralgias [abdominal pains or stomach aches] are here absolutely inefficacious, however zealously both physician and patient may employ them." This included the observation that anorexia involves a change in the body image: "the patient when told that she cannot live upon an amount of food that would not support a young infant, replies that it furnishes sufficient nourishment for her, adding that she is neither changed nor thinner."[9]

In the early part of the twentieth century, Wilder Penfield's mapping of the brain's somatosensory and motor areas, together with Henry Head's work on the body schema and Paul Schilder's conceptualization of the body image, represented major advances in our understanding of body representations. By 1950, the idea of the body image had pervaded neurology and psychiatry, and before long the concept was applied to the study of anorexia.

The German-American psychoanalyst Hilde Bruch was an early proponent of this body image approach. From decades of clinical experience, Bruch recognized that the primary characteristic of anorexia "is not the severity of the malnutrition *per se*, but rather the distortion of body image associated with it: the absence of concern about emaciation, even when advanced, and the vigor and stubbornness with which the often gruesome appearance is defended as normal and right, *not* too thin," Indeed, as she wrote in an influential 1962 paper, "This dramatic denial of illness alone—expressing a delusional disturbance in self-concept and body image—would justify considering these patients as suffering from a specific mental disorder." Importantly, Bruch also suggested that assessing the way anorexia patients think of their body is key to both diagnosis and treatment: "Evaluation of the disturbance in body image is of importance not only as a diagnostic criterion, but also in appraising treatment progress. Anorexic patients may gain weight for many reasons or may seem to progress well in psychotherapy. Without a corrective change in the body image, however, the improvement is apt to be only a temporary remission."[10]

Researchers used a number of methods to measure patients' perceptions of their body image, including questionnaires designed to assess body satisfaction, the "draw-a-person test," and the body image boundary score, based on the body boundary concept proposed by psychologist Seymour Fisher in 1964. Fisher believed that the robustness and malleability of the body boundary—the surface of the body, or the component of the body image that separates us from the outside world—differs from one person to the next, depending on personality, and assessed these properties from patients' verbal responses to an inkblot test. The resulting body image boundary score was based on responses that were broadly categorized as either "barrier imagery" or "penetration imagery" with the qualities of the patients' body boundary reflecting their attitude to their body.[11]

Beginning in the 1960s, researchers started to get more inventive, designing and building elaborate contraptions to measure body image perception more objectively. One device, built in 1962 by Donald J. Dillon, a psychiatrist at Columbia University, consisted of movable wooden beams forming something like a doorway. Standing six feet away from this structure, patients were asked to estimate their own body size—height, width, and depth—by moving the beams with ropes running through a system of pulleys to a revolving drum. At the end of the session, the patients would stand within the device, so that their estimate could be compared to their actual body size.[12] In 1976, a team at the Karolinska Hospital in Stockholm developed a system consisting of a video camera connected to a 26-inch TV monitor. Patients would sit in front of the camera, observing a distorted image of their body and had to adjust the image until it appeared accurate, using buttons on a control panel.[13] And, in 1964, a pair of psychologists in Chicago built an adjustable body-distorting mirror, which was bowed either way along its height and its width by motorized C-clamps attached to each edge. Patients stood seven feet away from the mirror and would see a series of grossly distorted reflections of their body, depending on the shape of the mirror: "tall, with pin head, large elongated body and legs tapering to tiny feet," "short with enormous horned head and tapering legs," "angularly distorted with one side of the body projecting laterally," and "a short body and dwarfish legs." Their task on each trial was to adjust the mirror until the reflection of their body seemed accurate. Initial tests, conducted on a sample of twenty hospital staff and ten psychiatry inpatients, had some unintended and undesirable consequences. A few of the study participants showed "exaggerated responses to viewing themselves in the mirror," including dizziness, nausea, and headache. Three of the psychiatry patients "adjusted the mirror so that they were grossly distorted in it," and one schizophrenia patient "was untestable because he fled from the testing room on seeing himself distorted in the mirror."[14]

Studies such as these produced mixed results. Some showed that patients with anorexia misjudge the dimensions of their body, significantly overestimating their body width and depth, in line with their deeply held conviction that their body is far larger in appearance than it really is. Other studies showed more modest overestimates, however, and still others showed no difference in estimates between anorexia patients and nonanorexic controls; indeed, that anorexics estimate the size of certain body parts more accurately than controls do, or that controls have a tendency to slightly overestimate the size of their body. These conflicting findings are likely due, at least in part, to the greater or lesser reliability of the methods used in the various studies to assess body image perception.[15]

Nevertheless, body image disturbance is today considered a core symptom of anorexia nervosa, and "disturbance in the way in which one's body weight or shape is experienced" is, according to the fifth edition of the *Diagnostic and Statistical Manual of Mental Disorders* (*DSM*), one of three criteria that must be met to reach a diagnosis of anorexia.* As Bruch suggested, it is also widely accepted that treatments targeting these disturbances are far more effective and long lasting than those which do not.

Early brain scanning studies showed that young and adolescent girls with early-onset anorexia exhibit reduced blood flow to a number of brain regions, including parts of the temporal and parietal lobes known to be involved in visuospatial abilities, and that women with anorexia have reduced gray matter volume in an area of the visual cortex involved in processing representations of body parts, compared to controls.[16]

Behavioral studies further suggest that people with anorexia have distorted body representations. For example, when study participants with anorexia are blindfolded and asked to judge the distance between two touch stimuli applied to various parts of their body, they tend to perceive the stimuli as being farther apart than they actually are, regardless of the sensitivity of the body parts that are touched.[17] Participants with anorexia also tend to overestimate the size—and especially the width—of their body in various laboratory tests. And these distortions affect their unconscious actions, too: when walking through narrow door-like openings, for example, they start turning far earlier than they need to, corresponding to openings 40 percent wider than their actual shoulder width.[18]

Thus unconscious body representations seem to be distorted, but the distortions filter through to conscious awareness, so that participants with anorexia perceive their body to be larger than it actually is. This negatively influences their attitude toward their body, causing an overall feeling of dissatisfaction that drives their unhealthy eating behavior.

Virtual reality offers promise to change the perception of, and attitude toward, their body for those with anorexia. In a 2014 study, researchers used a head-mounted display equipped with an optical motion-tracking system to induce the full body illusion in thirty-two nonanorexic female

*Lack of a menstrual period was removed from the diagnostic criteria for anorexia in the fifth edition of the *DSM* (2013) due to increasing recognition that the condition also affects young men.

participants. Synchronous touch applied to the participants' real and virtual body, along with congruent head movements, induced in them the illusion of ownership of a virtual body that was either much larger or much smaller than their own physical body. Those who took ownership of an underweight virtual body from a first-person perspective later experienced their own body to be smaller than it actually was, according to their responses on a questionnaire, and underestimated the size of their own body, whereas those who took ownership of an overweight virtual body experienced and estimated their own body to be larger.[19]

In a subsequent study, in 2016, researchers in the Netherlands used the same method to induce the full body illusion in thirty anorexia patients and twenty-nine control participants. In this case, however, the virtual body was an avatar with "neutral" dimensions—a waist-to-hip ratio of .75 and a waist circumference of about 72 centimeters (28 inches) considered by the World Health Organization to be "healthy." All of the participants estimated several different dimensions of their body size before the illusion, and, as expected, the anorexia patients overestimated the width and circumference of their shoulders, abdomen, and hips, but not their height. Afterward, the patients still overestimated the size of these same body parts, but less so than before the illusion. The control participants also perceived their body to be slightly smaller after the illusion, except for the circumference of their abdomen, estimates of which remained largely unchanged. The perceptual changes were larger for the circumference, than for the width, of these body parts, so the illusion seemed to make the participants feel their body to be less round. Importantly, the perceived changes in body size lasted for almost three hours after induction of the illusion.[20]

Body Dysmorphic Disorder

Most of us are unhappy about some feature of our body or appearance, and puberty is a particularly sensitive stage of life during which anxiety or embarrassment about our body is even more common. Adolescent angst usually disappears, but even as adults, many of us invest large amounts of time on activities that alter our looks and present a particular version of our self to the world—we apply makeup, wear a particular style of clothes, adhere to diets and go to the gym. Such activities become part of the daily routine for many millions of people around the world and take their place beside other aspects of their daily life. But an estimated 1 to 2 percent of people are preoccupied with how they look, becoming obsessed with one particular characteristic of their appearance.

Those with body dysmorphic disorder (BDD) have a persistent preoccupation with what they perceive to be a specific defect in their appearance. They can perceive such defects in any part of their body or aspect of their appearance. They may think, for example, that their hair is too thin, their breasts too small, or their buttocks too big. Or they may feel displeasure about their overall appearance, thinking, for example, that they are "fat and ugly." Most often, they are concerned with a particular facial feature—they may worry that their nose is too big, that their lips are too thin, that their jaw juts out, or that their skin is too pale. This causes a great deal of distress for those with BDD, who will typically perform certain behaviors over and over again, such as checking themselves in the mirror, grooming themselves excessively, or picking at their skin, and who will repeatedly compare their appearance to that of others. They may also seek reassurance to allay their concerns, but their "defects" are imperceptible, or at least insignificant, to other people, and so they will be told that they have nothing to worry about, or that they are making a fuss over nothing. But no amount of reassurance will stop them obsessing over their perceived defects.

BDD has a major impact on peoples' ability to lead a normal daily life. Those with BDD will often hide their secret obsession from the people they love, for fear of being dismissed, making it difficult, if not impossible, for them to maintain meaningful relationships. Some worry so much about their appearance that they will refuse to step outside, and they stay home from work regularly, so they may struggle to hold down a job. BDD is associated not only with high rates of social isolation and unemployment, but also with a high incidence of social anxiety, major depression, and even suicide.[21] A significant percentage of people with BDD request, and sometimes receive, multiple cosmetic surgical or dermatological procedures, but this rarely ameliorates their symptoms.[22]

The Italian psychiatrist Enrico Morselli accurately described BDD's cluster of symptoms in the 1880s, and named the condition dysmorphophobia, meaning "fear of ugliness": "The patient is really miserable; in the middle of his daily routines, conversations, while reading, during meals, in fact everywhere and at any time, is overcome by the fear of deformity . . . which may reach a painful intensity, even to the point of weeping and desperation."[23] The condition was given its modern name in 1987, and ten years later, researchers described a previously unrecognized form of BDD, which they named muscle dysphoria, and which is characterized by a pathological preoccupation with body size and muscle mass. People with muscle dysphoria become consumed by weight lifting and related activities to achieve their desired appearance—often including the use and abuse of anabolic steroids to help them build muscle mass.[24]

Like anorexia, BDD is considered to be a disturbance of the body image, typically associated with a feeling of unhappiness about the self and relationships with others. Little is known about the neurobiological basis of the condition, but research suggests that abnormalities in how the brain processes visual information may contribute to its symptoms. Functional neuroimaging studies show that the brains of people with BDD respond differently than those of controls to images of faces, and also exhibit reduced connectivity between frontal and occipital cortical areas involved in visual attention and deep brain structures that process emotional information. This widespread disorganization of neural networks may make them less able to integrate or organize disparate visual features into a meaningful whole; it could also help to explain the tendency of those with BDD to focus on specific facial characteristics, and the negative emotions they experience when doing so.[25]

Pain

Body image distortions are not restricted to rare neurological conditions and eating disorders but also occur far more widely. For example, chronic pain, defined as pain lasting three or more months, affects hundreds of millions of people and is a leading cause of disability worldwide. And chronic pain affects just about every part of the body. As "complex regional pain syndrome," it can be triggered by an injury and persist long after the injury has healed; as neuropathic pain, it may mysteriously appear without an obvious physical cause and is a feature of common painful conditions such as arthritis and carpal tunnel syndrome.

People experiencing arthritis, complex regional pain syndrome, or neuropathic pain, among other forms of pain, typically perceive the painful area to be increased in size, as one 2005 study reports,[26] and experiments using the "hand laterality test" show that this experience is associated with changes in the brain's representation of the body. In this test, volunteers experiencing pain and pain-free control volunteers are both shown images of hands in various positions and orientations and asked to judge whether a particular image is of a left or a right hand. When they see an image of a hand in an unnatural position, they have to imagine rotating the picture in their mind's eye before they can make the judgment. Even in pain-free controls, the time taken to judge laterality depends on the position of the hand in the image—the further the orientation of the hand in the picture is from their own hand, the more complex is the imagined movement, and the longer they take to respond. Thus motor imagery employs the brain's representations of the body, and these representations are subject to the same biomechanical laws that constrain movement of the body itself.

Patients experiencing complex regional pain syndrome (CRPS) perform significantly worse on the hand laterality test than non-CRPS controls, making fewer correct laterality judgments and also taking longer to make them than controls do. This depends, however, on the laterality of their pain—patients experiencing pain in their left arm show deficits in motor imagery of the left hand, but not of the right, and the opposite is true for those with pain in their right arm. In those experiencing pain, the ability to perform motor imagery on images of the unaffected hand is no different from the ability of the controls.[27] Brain scanning studies in patients with CRPS have arrived at conflicting results, however. Early studies using electroencephalography (EEG) and magnetoencephalography (MEG) showed the representation of the affected hand to be significantly shrunken, with the extent of shrinkage being closely related to the intensity of the pain felt in the affected hand (although these patients perceive the hand to be significantly larger than it actually is). But a more recent study, in 2019, using higher-resolution functional magnetic resonance imaging (fMRI) found that the cortical representations of the CRPS patients' affected hand, the corresponding hand of the controls, and the CRPS patients' unaffected hand were all comparable in size, location, and geometry.[28]

Carpal tunnel syndrome (CTS) occurs as a result of certain repetitive finger or wrist movements or of prolonged exposure to vibrations of the wrist, such as those produced by power tools. This places pressure on the median nerve as it passes through the carpal tunnel, an inch-wide passageway formed by the carpal bones of the wrist, causing pain, tingling, numbness, and weakness in the hand. In a 2012 study, researchers at Massachusetts General Hospital in Boston electrically stimulated the index, middle, and little finger of twelve CTS patients and twelve matched non-CTS controls. They measured the conduction velocity of the nerve impulses produced by the stimulation, and used MEG to examine the brain's responses to it. Impulses from the index finger and the middle finger, which are both supplied by the median nerve, were slower in the carpal tunnel syndrome patients than in the controls, whereas impulses from the little finger, supplied by the ulnar nerve, traveled at the same velocity in both groups. The slower conduction velocity of median nerve impulses in the CTS patients corresponded to delayed responses in the somatosensory cortex, with the extent of the delay being related to the severity of the pain these patients were experiencing. Furthermore, the responses to index and middle finger stimulation were larger in the CTS patients than in the controls, and not as distinct from each other, suggesting that cortical representations of the fingers had become blurred in the CTS patients.[29]

A larger follow-up study in 2014, also at Mass General, confirmed that these functional differences reflect reorganization of the hand map in the somatosensory cortex. This time, the researchers applied vibrating touch stimuli to the same three fingers in sixty-three CTS patients and twenty-eight matched controls, and used fMRI to examine the brain responses. They found that the distance between somatosensory cortical representations of the index finger and the middle finger was smaller in the CTS patients than in the controls. The ability to pinch with these two fingers was weaker and slower and the ability to discriminate between two touches on those fingers was also weaker in the CTS patients than in the controls.[30]

Anesthesia

Unpleasant as it may be, pain serves the vital evolutionary purpose of drawing our attention to potentially life-threatening injuries. If bodily awareness usually lies at the fringes of consciousness, then the body image distortions caused by pain may help draw our attention to an injury by making us *hyper*aware of the injured part of the body. But the drugs we use to treat pain can also distort the way we perceive our body; and our growing knowledge of the brain mechanisms underlying bodily awareness is providing new and surprisingly simple ways of alleviating the pain we feel under certain circumstances.

Anesthetics are a marvel of modern medicine used to numb specific parts of the body or to induce total loss of conscious awareness. Anesthesia means "loss of sensation," and these drugs act by blocking the nerves that carry sensory information from the body into the spinal cord and then up into the brain, but they can also block the motor nerves, resulting in muscle relaxation or paralysis, and they can sometimes also cause memory loss. Anesthetics are administered widely and routinely, to relieve the pain of an injury or to prevent it altogether during a surgical procedure. As anyone who has received a local injection of lidocaine at the dentist's will know, anesthesia can distort the way we perceive our body, and this has been studied both under laboratory conditions and in patients on the operating table.

In a 1999 study by researchers at the Prince of Wales Medical Research Institute in Sydney, Australia, one set of participants had one of their thumbs completely anesthetized with an injection of lidocaine, and another set had their lips numbed with anesthetic cream. Before and during anesthesia, the participants were shown template sheets containing drawings of thumbs and lips of varying size, and repeatedly asked to choose the drawing that best matched the perceived size of their own thumb or lips. Anesthesia increased participants' perceived size of the thumb by 60 to 70 percent, and that of the lips by around 50 percent. In another experiment, a third group of participants drew outlines of their hands or lips while blindfolded before and during anesthesia. Again, anesthesia increased the participants' perceived size of their lips and their thumb—but not of their entire hand—showing that the perceptual distortions influenced the function of the motor system. Thus the body image distortions produced by anesthesia are accompanied by distortions of voluntary actions.[31]

In 2003, clinicians at Pitié-Salpêtrière University Hospital in Paris studied body image distortions produced by local anesthesia in thirty-six patients undergoing orthopedic surgery on a locally anesthetized arm or leg. All but six of the patients described perceptual distortions of the size of the anesthetized limb. The others all reported the perception that the width of the anesthetized limb had increased, compared to the limb on the other side of their body, the perception of which remained unchanged, and some patients also said that the anesthetized limb seemed longer or shorter.

These perceptual changes were dramatic but varied rather widely, depending on the exact location at which anesthesia was administered. Some of the patients stated that the size of the anesthetized limb had increased two- or threefold, and one patient reported that his legs felt "immense"; several said that the perceived width increase was more vivid in their hand or foot; and another said that his arm, but not his hand, felt enlarged. Those who received two sequential anesthetics in different leg nerves reported changes in the perceptual distortions: blocking of the sciatic nerve produced the sensation that the outer half of the leg, but not the inner, felt wider below the knee, and subsequent blocking of the femoral nerve or its saphenous branch produced the feeling that the whole leg was swollen. And four male patients under spinal anesthesia felt that the size of their penis had increased.

Importantly, the patients reported feeling that their anesthetized limb began to swell up following administration of the anesthetic, but they did not become fully aware of the perceptual distortion until prompted to focus their attention on the size and shape of the anesthetized limb. When they did become aware of the changes, they noticed the discrepancy between their perception of the anesthetized limb and its actual size, and looking at or grasping the limb made them realize that the perceived changes were illusory, and they corrected their perception of it. But two of the patients completely failed to recognize the anesthetized limb as their own when they looked at it, telling the researchers that "this is not my arm," and they looked for it in the wrong place. This "swelling illusion" begins within minutes and seems to predict both the spread and the effectiveness of the local anesthetic very accurately.[32]

Following up on these findings, in 2011, a research team at the University Hospital of Toulouse asked surgical patients to perform the hand laterality judgment task during anesthetic blocking of the brachial plexus, the network of nerves that extends through the shoulder and out along the arm, with their anesthetized arm hidden behind a curtain. The patients' judgments of images corresponding to their anesthetized hand were less accurate, and took significantly longer, than those of images corresponding to their other hand. Their performance improved somewhat, however, when the curtain was pulled back so they could see their anesthetized arm.[33]

This study is one of many that highlight the importance of the sense of sight for bodily awareness, and a number of different research groups have found that simple illusions that manipulate how we see our body can have an analgesic effect. Thus, for example, a 2008 study found that in patients with chronic limb pain, the pain produced by movement intensifies when they see a magnified view of the affected limb through binoculars, and the swelling of the limb also increases. By contrast, when the binoculars are turned around to make the limb look smaller, the pain they feel and the swelling of the limb are both reduced. Magnifying the view of the limb also increases its sensitivity to touch, while diminishing the view of it has the opposite effect.[34]

In another 2011 study, similar effects were seen in osteoarthritis patients who viewed real-time video footage of their arthritic hand on a television screen. Manipulating the image to alter the appearance of the painful part of the hand reduced the amount of pain the patients felt. Some said shrinking the image reduced their pain, others said stretching it was beneficial, and a few said that both shrinking and stretching the image were beneficial. The pain was reduced significantly in all the study participants and temporarily eliminated in one-third of them. Manipulating the image of unpainful parts of the arthritic hand or of the unaffected hand had no effect.[35] Exactly why visual perception has such a powerful effect is unclear, but one possibility is that altering the perceived size or shape of the affected limb or body part reduces the sense of ownership over it.

Other studies show that, by merely looking at their body, participants can reduce the acute pain arising from tissue damage caused by an injury. In a 2009 study, researchers applied an infrared laser beam to uninjured participants' right hand. The intensity of the pain they felt decreased when they looked directly at their right hand, or at a reflection of their left hand seen in the location where they felt their right hand to be, but looking at the reflection of somebody else's hand, or of some other object, in the place of their right hand had no such effect.[36] Brain scanning in a 2012 study further revealed that the reduction in pain participants felt while looking at their injured hand is associated with reduced pain-related activity in their somatosensory cortex, and with greater communication between the pain-processing areas and those parts of the parietal lobe thought to contain the higher-order representations of the body. Thus it appears that the sense of sight exerts "top-down" effects, whereby the parietal regions encoding higher-order body representations can influence earlier stages of sensory processing.[37]

Drugs

Various narcotic drugs can profoundly distort the body image, with hallucinogens being a well-known example. The Slovenian psychiatrist Alfred Serko published one of the earliest scientific accounts of such effects, after experimenting with mescaline, in 1913: "I felt my body particularly plastic and minutely carved. At once I had a sensation as if my foot was being taken off . . . then I felt as if my head had been turned by 180 degrees. My abdomen became a fluid, soft mass, my face acquired dimensions, my lips swelled, my arms became wooden, my feet turned spirals and scrolls, my jaw was a hook and my chest seemed to melt away."[38]

Apparently innocuous medicinal drugs can have similar effects if taken in sufficient quantities. In 1999, for example, a team of Japanese psychiatrists reported the case of Mr. A., a forty-six-year-old man who developed Alice in Wonderland syndrome following a prolonged period of cough syrup abuse. Although the man did not have a history of psychiatric illness and did not drink alcohol regularly, he did take three different blood pressure–lowering drugs daily. What's more, he had also consumed almost ten times the recommended dose of Dickinin Cough Syrup regularly for over three years, after which time the strange experiences began: books and TV images seemed to be distorted, his body seemed to expand, insects appeared to be enlarged, and he began to feel afraid and to experience insomnia. Doctors prescribed an antianxiety drug, but five days later, he overdosed on it, became confused, and stabbed himself. Unable to explain his behavior, Mr. A. was hospitalized, and ordered to stop taking the cough syrup immediately. He was also given an antipsychotic drug and the anti-insomnia ("date rape") drug Rohypnol to help him sleep. His symptoms persisted for a week or two—he described seeing "a lot of small men and women in colorful clothes" walking around the hospital ward—but they gradually diminished and then disappeared altogether within a month of his admission to the hospital.[39]

8 Interoception

Bodily awareness depends on a continuous stream of signals from the body to the brain. Broadly speaking, the brain receives three types of sensory signals—exteroceptive, interoceptive, and proprioceptive. Exteroceptive signals arise from outside the body, relayed from the outside world by the sense organs: visual signals from our eyes, auditory signals from our ears, and the various types of touch signals from our skin. Interoceptive signals arise from inside the body and are related to its physiological state;[1]* they include messages from our gut, heart, lungs, and hormonal system. They enter our consciousness to alert us to some physiological need—we experience feelings of hunger, thirst, and a full bladder, for example. Finally, there are proprioceptive signals, which make us aware of the position and movement of our limbs. These also arise from inside the body and are sent from stretch receptors in our muscles, tendons, and joints. As with exteroceptive signals, some interoceptive signals and some proprioceptive signals enter our conscious awareness, but others do not. Nevertheless, even those we are unaware of exert subtle but significant influences on our thoughts and behaviors in ways we are only just beginning to understand, and they play an important role in bodily self-consciousness.

The adult human brain is an extremely hungry organ that consumes approximately one-quarter of the body's oxygen intake, despite accounting for just one-tenth of total body weight. Physicians often tell their patients

* Most of the physiological functions that generate interoceptive signals are controlled autonomously. We know, however, that Tibetan Buddhist monks experienced in transcendental meditation can consciously control their own heart rate and body temperature, and researchers recently reported that some individuals can consciously generate goosebumps.

that "what's good for the heart is good for the brain," meaning that the lifestyle choices considered beneficial for cardiovascular health—such as exercising regularly, eating a balanced diet, and not smoking—are also beneficial for brain health, and reduce the risk of stroke, dementia, and other neurological problems. Furthermore, it is well established that aerobic exercise—or lack thereof—can significantly impact mood.

The heart influences the brain more directly, however. Numerous magnetic resonance imaging studies published in the past decade or so show that the brain pulsates in time with the beating of the heart, rhythmically bulging and shrinking as blood is pumped into it and then drained out. These movements are miniscule—approximately 50 microns (thousandths of a millimeter) or less, smaller than the width of a human hair—but are sufficient to displace electrodes placed deep into the living brain by about 3 microns, causing minor variations in their recordings. The heart literally moves the brain, and this "cardioballistic" motion can influence the behavior of individual neurons.[2]

The heart speaks to the brain by means of pressure and stretch receptors in the aorta, the large artery that carries blood around the body. Every time the heart muscles contract to pump blood out (a phase of the cardiac cycle called the systole), these receptors detect the pressure changes in various locations near the heart and signal the strength and timing of each heartbeat to the brain. The brain responds, if necessary, by altering the heart rate and directing the release of hormones that either constrict or dilate the blood vessels to regulate blood pressure. We are usually unaware of our heartbeats, but we can perceive them if we concentrate on doing so, and we become fully conscious of our heart beating when we are angry or afraid. Laboratory studies of interoception commonly use the heartbeat perception task, in which participants are asked to count their heartbeats for a minute or two, without feeling their pulse; their estimates are then compared to the actual number of heartbeats, as determined by a fingertip pulse oximeter, or some other heart rate measuring device. Heartbeat sensitivity differs widely between individuals, and this is reflected in the accuracy of the estimates they give; but research also suggests that training can improve an individual's accuracy on this particular task.

The heart's activity has a subtle effect on our perceptions and thoughts. According to research published in 2013, the cardiac cycle can interfere with our ability to attend to words and commit them to memory. The researchers adapted a method used to test a phenomenon known as emotional attentional blink, whereby one visual stimulus (such as a word flashed on a screen) "masks" the detection of a second stimulus shown fractions of a second afterward; if, however, the second stimulus has high emotional content, it can break through the "mask" and be detected. In this 2013 study, presentation of the second word in each pair was timed so that it appeared during either the systole or the diastole (the other phase of the cardiac cycle, during which the heart muscles relax so that blood can enter). Participants' memory of the words they had seen was tested, and the results confirmed that emotional words presented are more likely to break through the attentional blink. Crucially, though, breakthrough was more likely for those words flashed during the diastole than during the systole. Furthermore, participants with higher interoceptive sensitivity, as measured by the heartbeat detection task, were less susceptible to the interfering effects of the heartbeat and recalled more of the words presented during systole.[3]

The cardiac cycle also influences more complex psychological processes, such as negative racial stereotyping. In the Western world, Black people may be perceived as threatening by non-Blacks, and so harmless objects they are holding are more likely to be misidentified as weapons. This implicit bias can influence behavior; it helps to explain racial bias in police shootings in the United States. In the lab, perception of racial stimuli appears to be modulated by signals from the heart to the brain. In one experiment of a 2017 study, researchers showed photographs of Black and White people's faces alongside images of tools or handguns and asked their non-Black participants to identify the object they had seen. They found that participants were more likely to misidentify the tools as handguns when the images were presented during the systole than during the diastole. In a first-person shooter game, Black and White characters appeared on the screen, and the participants had to quickly decide whether or not to shoot, based on whether the character was holding a gun or some other object. Overall, they were more likely to shoot unarmed Black characters than White ones, but their tendency to do so was greater when the characters appeared during systole than during diastole. Interestingly, though, cardiac signals apparently do not affect the neural processing of information associated with positive racial stereotypes—in another experiment of the 2017 study, when participants viewed images of Black and

White athletes paired with either fruits or sports-related objects, the cardiac phase timing of the images did not affect their ability to identify the objects.[4]

Various other physiological functions are also coupled to the heartbeat. For example, one 2016 study found that microsaccades, the small, involuntary eye movements that fixate gaze onto a single location, occur far more frequently during the systole than during the diastole phase of the cardiac cycle. This could be due to the effects of the heartbeat on visual processing, or because of direct effects of the heartbeat on areas of the brain stem that control eye movements.[5] Breathing—which is closely linked to the cardiac cycle—also influences brain function: it synchronizes the electrical activity of neuronal populations in deep brain structures and various areas of the cerebral cortex, such that, as another 2016 study found, memory recall and emotional recognition are more accurate during breathing in than breathing out.[6]

Interoception and Bodily Self-Consciousness

Interoceptive signals interact with the exteroceptive signals underlying our sense of body ownership. The less accurately study participants perform the heartbeat perception task, the more susceptible they are to the rubber hand illusion. That is, participants with low interoceptive accuracy experience the illusion more robustly and feel a stronger sense of ownership over the fake hand than those with high interoceptive accuracy do. A variation of the rubber hand illusion shows that cardiac feedback can alter participants' experience of body ownership. The cardiac rubber hand illusion can be induced with a head-mounted display that projects a photo-realistic virtual rubber hand into the visual field. In one experiment of a 2013 study, published in the journal *Neuropsychologia*, this illusion is induced by touch applied synchronously to the real and virtual hand, but the appearance of the virtual rubber hand is altered by the cardiac feedback to make it flicker, or appear to throb, either in or out of time with the heartbeat. Participants' sense of ownership over the virtual rubber hand was found to increase when the hand flickered in time with the heartbeat, and to decrease when it flickered out of time. In another experiment in that study, the head-mounted display showed a real-time video silhouetted image of an entire body. Participants' sense of ownership over the virtual body was found to increase, and their touch sensations more likely to be mislocalized toward it, when the silhouette was lit in synch with the heartbeat. This cardiac full body illusion did not occur when the virtual body was lit out of synch with the heartbeat, however.[7]

Thus, the brain integrates signals from both inside and outside the body to generate our sense of bodily self-consciousness, and our sensitivity to signals arising from inside the body predicts the malleability of our body representations, and our susceptibility to multisensory illusions. As a 2018 study shows, the heart's activity has a direct influence on the processing of touch-related information in the somatosensory cortex, and on conscious perception of it. In this study, researchers used scalp electrodes to record two distinct early responses in the somatosensory cortex during the cardiac full body illusion, approximately 20 milliseconds apart. They found an increase in participants' sense of ownership over the virtual body when it lights up in time with their heartbeat, and that this is correlated with an increase in their perceived size of the virtual body. The size of several later somatosensory cortical responses to touch decreases when the stimuli coincide with the systole, and these touch stimuli are less likely to enter participants' awareness. Participants are also more likely to perceive small electrical pulses applied to an index finger as touch when these are applied during the diastole than during the systole, and those pulses that are consciously detected are quickly followed by a significant drop in the participants' heart rate, whereas those that are not detected are followed by a significant rise.[8]

Signals from the heart influence the sense of agency, too. In a 2020 study, when asked to choose between pairs of playing cards to win or lose money, participants felt a greater sense of control over their action (pressing a button) during the systole than during the diastole, regardless of the outcome.* In this study, the researchers measured agency by participants' explicit subjective reports of the extent to which they felt in control, and also by their judgments of the time interval between action and outcome (a measure of intentional binding). Even though participants reported a reduced sense of agency for decisions that resulted in losing money, they

* Interestingly, other research suggests that study participants more often perform voluntary actions while breathing out than when breathing in. The participants in these 2020 experiments showed no such preference, however.

still perceived the interval between their pressing a button and its outcome to be slightly shorter than it actually was.[9]

The Insula

The brain contains a network of structures that process internal bodily signals, a key component of this interoceptive system being the insula, which lies deep inside the lateral sulcus separating the temporal and parietal lobes of the brain. The insula is an evolutionarily recent structure, which is found only in primates and is most prominent in humans. It receives sensory inputs from the body via the brain stem and thalamus and is divided into distinct nuclei, or subregions, which process bodily signals sequentially, with information flowing through the insula from its back to its front. The posterior (back) region encodes a "primary" interoceptive representation, which maps bodily feelings such as pain, temperature, itch, visceral sensations, hunger, and thirst at high resolution; the mid-insula integrates bodily feelings with other sensory information that gives those feelings emotional meaning; the anterior (front) region brings this information into conscious awareness and also relays it to the anterior cingulate cortex.[10]

Both the insula and cingulate cortex are activated during the heartbeat perception task, and researchers have found that activity of the insula is closely related to performance on the task, with those performing better showing greater activity in the insula; in other words, stronger representation of cardiac signals in the brain is closely related to greater awareness of the signals. The insula is also sensitive to signals from outside the body and processes them regardless of conscious awareness. In brain scanning experiments, the exact timing of visual stimuli determines whether or not they enter participants' awareness, and this is closely linked to the level of activity in the insula. Images presented at the same time as the heartbeat take longer to enter visual awareness, or may not enter it at all, and yet the insula still responds to these images albeit less so than to images presented at other times. This may be because the brain has to monitor the heartbeat while also keeping its effects on sensory perception at a minimum. The high level of cardiac arousal during the systole interferes with the processing of sensory information, so the brain may actively suppress cardiac signals to reduce this interference. Thus, as well as encoding representations of interoceptive signals and bringing them into conscious awareness, the insula also integrates these with exteroceptive signals and regulates their access to awareness, too.[11]

Although the word "insula" means "island," the structure does not act alone. The insula and anterior cingulate cortex are the main nodes of a network of brain structures that process interoceptive information, and this network is one component of a larger, brain-wide network that couples sensations from inside and outside the body. As a 2019 study shows, this larger network, which includes the primary motor and somatosensory cortices and "association areas" of the frontal, temporal, and parietal lobes of the brain, processes and integrates interoceptive and exteroceptive signals to generate the building blocks of self-consciousness—our senses of agency and body ownership.[12]

Emotions

Every waking moment we experience a continuous and changing stream of signals from inside the body. Different mental and emotional states are associated with particular inner sensations or physiological responses, only some of which enter our conscious awareness. Some of these sensations are easy to identify and cause us to act upon them—we seek food when we feel hungry, for example, and drink when we feel thirsty, but others may be more difficult to interpret. Nevertheless, our feelings are closely associated with particular bodily sensations, and both are almost always linked to an emotion of some kind.

Researchers in Finland surveyed more than 1,000 people in 2018 to assess the bodily sensations and mental experiences that accompany 100 "core" subjective feelings, including positive emotions such as happiness and pride, negative emotions such as fear and anger, and feelings associated with cognitive functions (such as imagining, reasoning, and remembering), illness (dizziness, coughing, sneezing), and homeostasis (hunger, thirst, breathing), to map what they call the "human feeling space," showing how different feelings and emotions are linked to specific sets of bodily sensations. Thus feeling angry is associated mostly with intense sensations in the head and hands; anxiety with sensations in the chest; and relaxation with sensations of moderate intensity all over the body.[13]

Different feelings or emotions can be associated with similar or overlapping bodily sensations. For example, we often say that we have "butterflies in our stomach" when we are anxious or excited. This feeling occurs when the blood vessels around the stomach and intestines constrict, reducing blood flow to these organs. It is part of the "fight or flight" response, a set of automatic physiological mechanisms that occur when we are in danger, which is triggered by hormones in our blood, and which diverts blood toward the muscles and legs. The way we interpret this fluttering feeling depends on the situation we are in when we experience it.

William James recognized the link between interoception and emotion more than a century ago. "Our natural way of thinking about these standard emotions is that the mental perception of some fact excites the mental affection called the emotion, and that this latter state of mind gives rise to the bodily expression," he wrote in1884. "Common sense says, we lose our fortune, are sorry and weep; we meet a bear, are frightened and run; we are insulted by a rival, are angry and strike." But James turned conventional thinking about our emotional experiences on its head, suggesting instead that this sequence of events occurs in reverse:

> My thesis on the contrary is that *the bodily changes follow directly the* PERCEPTION *of the exciting fact, and that our feeling of the same changes as they occur* IS *the emotion.* . . . The hypothesis here to be defended says . . . that the one mental state is not immediately induced by the other, that bodily manifestations must first be interposed between, and that the more rational statement is that we feel sorry because we cry, angry because we strike, afraid because we tremble and not that we cry, strike, or tremble, because we are sorry, angry, or fearful, as the case may be. Without the bodily states following on the perception, the latter would be purely cognitive in form, pale, colourless, destitute of emotional warmth. We might then see the bear, and judge it best to run, receive the insult and deem it right to strike, but we could not actually *feel* afraid or angry.[14]

In recent years, researchers have reformulated this idea as the "active interoceptive inference" hypothesis, according to which the brain makes predictions about changes in bodily states and uses these predictions to control the body's reflexive physiological responses to events in the outside world. The predictions are compared to the physiological changes that actually occur, and any discrepancies are used to update future predictions and make them more accurate. These predictions serve, first and foremost, to regulate the body's energy consumption efficiently, by anticipating its energy requirements and meeting those needs before they arise. Thus the brain preempts thoughts of starting to run by increasing the heart rate and redistributing the flow of blood from organs such as the stomach to the leg muscles, and the associated sensations join the interoceptive signals arising from the body. Our conscious emotional experiences are little more than a by-product of these processes, ascribed after the fact by our beliefs about the causes of internal bodily signals.[15]

People differ widely in their sensitivity to internal bodily signals and in their ability to identify them accurately. Those with high interoceptive accuracy may be better able to direct their attention toward interoceptive signals and bring them into conscious awareness when necessary. Their brains predict signals arising from inside more precisely, and they ascribe feelings to these sensations more readily. By contrast, those with lower interoceptive accuracy make less precise predictions about these signals, find it harder to make sense of the sensations, and are more susceptible to false beliefs about them. Indeed, study participants who score high on the heartbeat perception task experience bigger changes in heart rate when viewing pleasant and unpleasant images, report greater levels of emotional arousal in response to the images, and identify the emotions induced by them more easily. They can also describe the bodily sensations induced by emotional stimuli more specifically.[16]

The role of attention in active interoceptive inference may explain why we do not normally perceive our heartbeats. The heartbeat is usually stable, and as long as it remains predictable, it goes unnoticed. But unexpected changes in the strength or rhythm of our heartbeats enter conscious awareness because we may need to act upon those changes. It may also explain why people's performance on the heartbeat perception task varies so widely. Participants who can attend closely to their interoceptive signals therefore predict their heartbeats more accurately, without necessarily being conscious of them.[17]

Clinically, difficulty in identifying and describing felt emotions is referred to as alexythymia or "emotional blindness" and was traditionally ascribed to an impaired ability to perceive the internal bodily signals associated with emotions. Recent research suggests, however, that alexythymia is not specific to emotional states, but extends to nonemotional signals, too. It also shows that people with alexythymia not only have difficulty in interpreting their emotions, but they also perceive their own heartbeats less accurately than others; moreover, they tend to consume more coffee and alcohol and to take longer to seek medical attention after having a heart attack. In their responses on questionnaires, alexythymia respondents report that they often forget to eat, or that they eat until they feel uncomfortable; that they are rather clumsy and uncoordinated, and often find cuts and bruises that they cannot explain; and that they do not know how firmly they should hug others. They are also more likely to confuse their own emotional states with feelings such as hunger, thirst, and arousal. Thus those with alexythymia seem to be less sensitive to physiological signals about the state of their body or, at least, are more likely to misinterpret their bodily signals than others.[18]

Alexythymia may therefore be more accurately defined as "a general deficit in interoception." It is associated with abnormalities in the structure and function of the anterior insula and anterior cingulate cortex, and occurs along with a variety of clinical disorders known to involve poor interoception. One such disorder is anorexia nervosa. The pioneering psychoanalyst Hilde Bruch recognized that disturbances of interoception play an important role in anorexia. "The second outstanding characteristic of the anorexic patient is a disturbance in the accuracy of perception or cognitive interpretation of stimuli arising in the body," she wrote in her seminal 1962 paper, "with failure to recognize signs of nutritional need as the most prominent deficiency of this type." This disturbance, she continues, is "more akin to inability to recognize hunger than to mere loss of appetite. . . . Awareness of hunger and appetite in the ordinary sense seems to be absent. . . . Even though occasionally hunger may become overwhelming in its biological urgency, usually there is denial and non-recognition of the pains of hunger, even in the presence of stomach contraction. On the other hand, there are complaints of acute discomfort and fullness after the intake of the smallest amount of food."[19]

As yet, there are very few behavioral studies investigating interoception in patients with anorexia, but the available research suggests that the condition does indeed involve impaired perception of internal bodily signals. Patients with anorexia perform poorly on the heartbeat perception task, compared to nonanorexic controls, and report difficulties in recognizing sensations related to hunger. These results point to a disturbance in interoceptive awareness, whereby sensitivity to visceral signals is reduced. On the other hand, sensitivity to other internal signals may be heightened. In one 2015 study, with fifteen female anorexic participants and fifteen matched controls, researchers gave all thirty participants the heart rate–increasing drug isoproterenol and tested their perception of cardiac and breathing-related signals before and after they ate a meal. Surprisingly, the anorexic participants were more likely than the controls to report feeling an increased heart rate and shortness of breath, even when given the lowest dose of the drug or a sugar solution placebo, before eating the meal. Thus, because they found it difficult to separate predicted interoceptive signals from actual ones and perceived certain of these signals more intensely in some contexts, the anorexic participants experienced a state of heightened physiological arousal—or anxiety—in anticipation of eating.[20]

Brain scanning experiments provide evidence that anorexia and other eating disorders involve abnormal processing of interoceptive signals, particularly those related to food and eating. For example, study participants who have recovered from anorexia display increased activation of their insula when viewing images of food compared to controls, which may contribute to feelings of anxiety toward food and eating. They also exhibit significantly lower insula activity when tasting food and water than non-anorexics do, suggesting that they process taste sensations differently.[21]

Individual differences in the ability to sense and identify internal bodily signals may contribute to numerous other conditions and behaviors. Anxiety and depression, for example, may involve a disconnection between events in the outside world and the interoceptive signals they produce. Normally, our subjective experiences alter the milieu of internal bodily signals temporarily, and our perception of these changing bodily states plays a role in our interpretation of, and beliefs about, our experiences. These processes seem to go awry in those with high trait anxiety, such that the physiological state of their body does not quickly revert to baseline, causing them to attribute too much emotional significance to external events. Similarly, some people experience hypochondria, or pathological anxiety about their health, as a result of their increased sensitivity to interoceptive signals, inability to interpret them, or both. On the other hand, depression seems to be associated with blunted physiological responses to pleasurable experiences, or at least with a reduced sensitivity to them, and those who self-harm and attempt suicide exhibit interoceptive "numbing," characterized by a higher tolerance for unpleasant sensations.[22]

9 Add-Ons

Some thirty inches from my nose
The frontier of my Person goes,
And all the untilled air between
Is private *pagus* or demesne.
Stranger, unless with bedroom eyes
I beckon you to fraternize,
Beware of rudely crossing it:
I have no gun, but I can spit.
—W. H. Auden, "I Have No Gun, but I Can Spit"

Our bodily awareness extends beyond the physical boundaries of our body to the space immediately surrounding it—our "peripersonal space," within which we can reach and grasp objects. Early investigators speculated that the body schema and body image extend to incorporate foreign objects such as clothes and tools. Modern research confirms this, and reveals that tool use alters body representations. Recent advances in our understanding of bodily awareness are now enabling researchers to build sophisticated prosthetic limbs that interact with the brain and feel like a real part of the amputee's body rather than a cumbersome attachment.

Each of us inhabits the volume of space that contains our body, but we also feel a sense of ownership over the space immediately surrounding our body, a "personal space" that forms a boundary between our self and others. In public places, we follow unspoken rules about these invisible boundaries, and consistently maintain a distance to separate our self from others.

Anywhere other than a crowded place, the closeness of a stranger makes us uncomfortable—by encroaching upon our personal space, the stranger is invading our privacy.

In his 1966 book *The Hidden Dimension*, the American anthropologist Edward T. Hall defined "personal distance" as "a small protective sphere or bubble that an organism maintains between itself and others," the innermost of "a series of invisible bubbles," or "distance zones" that surround us. Distance zones have measurable dimensions, and each one evokes unique multisensory experiences. "Public distance" (12 feet or more) is "well outside the circle of involvement, and it is only inside the three inner zones that interpersonal interactions take place. Formal business and social discourse take place at the far end of "social distance" (7 to 12 feet), at which "skin texture, hair, condition of teeth, and condition of clothes are all readily visible . . . and the eyes and mouth of the other person are seen in the area of sharpest vision." In close social distance (4 to 7 feet), "the area of sharp focus extends to the nose and parts of both eyes; or the whole mouth, one eye, and the nose are sharply seen." This is the distance at which impersonal business occurs . . . . People who work together tend to use close social distance. It is also very common for people who are attending a casual social gathering."

Hall puts the far end of "personal distance" at 4 feet—at "arm's length," or "just outside easy touching distance by one person to a point where two people can touch fingers if they extend both arms." At this distance, "head size is perceived as normal . . . and breath odor can sometimes be detected." At the close end of personal distance (1 to 2.5 feet), "one can grasp the other person," and "there is visible feedback from the muscles that control the eyes." "A wife can stay inside the circle of her husband's close personal zone with impunity," Hall writes, but "for another woman to do so is an entirely different story."

At the far end of "intimate distance" (6 to 18 inches), "heat from the other person's body, sound, smell and feel of the breath all combine to signal the unmistakable involvement with another body," and so, standing within this distance of another "is not considered proper by adult, middle-class Americans." Although public transportation "may bring strangers into what would ordinarily be classed as intimate spatial relations . . . subway riders have defensive devices which take the real intimacy out of intimate space in public conveyances." The basic tactic commuters use to avoid intimate touch, Hall writes, "is to be as immobile as possible and, when part of the trunk or extremities touches another person, withdraw if possible. If this is not possible, the muscles in the affected areas are kept tense." Closer yet is the inner intimate distance (6 inches or less). This is "the distance of love-making and wrestling, comforting and protecting," at which "physical contact or the high possibility of physical involvement is uppermost in the awareness of both persons." At this distance, sharp vision is blurred, but the smell and body heat of the other is easily detected; speech "plays a very minor part in the communication process," and "a whisper has the effect of expanding the distance."[1]

Hall thus linked distance zones to patterns of behavior, noting that "where people stand in relation to each another signals their relationship, or how they feel toward each other, or both." He also noted how the silent language of body spacing—which he named "proxemics"—differs between cultures, and devoted a section of his book to the "proxemic patterns" of the Arabs, Europeans, and Japanese, and how they differ from those of Americans. These "manifold unstated, unformulated differences," he wrote, "often result in the distortion of meaning, regardless of good intentions, when peoples of different cultures interact."[2]

Hall was most interested in cross-cultural business interactions, and in how the way people think about and use the space around them may inform architecture and urban design. Crucially, though, he recognized that "the boundaries of the self extend beyond the body," and that "man's sense of space is closely related to his sense of self," ideas that neurologists had entertained since the early part of the twentieth century, and which have now been confirmed by modern brain research.[3]

Peripersonal Space

Studies performed on macaques in the 1980s began to reveal that various parts of the monkey brain contain cells that respond to different kinds of sensory stimuli outside the body. In one 1981 study, Italian researchers used microelectrodes to record the activity of seventy-two "mouth" neurons in the premotor cortex of three pigtail macaques. All of the cells responded to some kind of sensory stimulation, in one of three ways. Cells of one type responded strongly to passive touch on the mouth or the skin surrounding it; cells of another type responded to the same passive touch stimuli, but markedly stepped up their activity if the monkey performed any kind of mouth movement immediately after the touch stimulation; and cells of a third type responded strongly only when touch stimulation was accompanied by movement. More than half of the cells also responded to visual stimulation; some responded to visual stimuli alone, and others became more active when the monkey performed a specific mouth movement after visual stimulation.[4]

Subsequent studies in 1994 and 1997 showed that this same region in the monkey brain contains large numbers of neurons that respond to visual stimuli in the space immediately surrounding the arm and hand, regardless of limb position, and independently of eye movements; that the cells are arranged somatotopically, with adjacent cells responding to stimuli on or near adjacent parts of the body; and that subsets of the cells respond to objects near the body, and continue to encode the objects' location even in the dark after they have been removed. Together, these findings show that the premotor cortex of macaques contains neurons that respond to both touch and visual stimuli, and also encode maps of the body and of the area immediately surrounding it. These cells likely integrate sensory and motor information before transmitting signals to the motor cortex and other regions involved in executing movements. Cells with these same properties have also been found in various other brain regions including, notably, the parietal lobe, where higher-order body representations are encoded.[5]

Experiments employing an unusual stimulation method have since revealed an unexpected organizing principle, however. Electrical stimulation of the monkey motor cortex is a relatively common research technique that has been used since the late nineteenth century. Typically, this stimulation was applied to the motor cortex on one side of the brain in brief pulses lasting no longer than 50 milliseconds, enough to elicit twitches in the corresponding muscle groups on the opposite side of the body without spreading to connected brain regions. In a 2016 study, however, one group of researchers tried stimulating the macaque premotor cortex for longer periods of time, and found that extending the stimulation for about 500 milliseconds elicited complex, meaningful movements.

Furthermore, stimulation of different zones within the premotor cortex evoked different types of complex movements. Stimulation of one zone in a given hemisphere produced hand-to-mouth movements on the opposite side, in which the monkey's mouth opens, its hand assumes a precision grip posture, and its forearm, elbow, shoulder, and neck rotate, bringing its hand toward its mouth; stimulation of an adjacent zone evoked various mouth movements, such as chewing and licking, of another zone produced climbing and jumping movements, and of yet another zone elicited what appeared to be defensive movements, such as closing the eye, turning the head away, and lifting the hand to its face. Importantly, the zones producing different types of movements were found to be arranged topographically across the motor and premotor cortices, and stimulation of a specific zone produced exactly the same coordinated movement, with the limb on the opposite side ending in an identical location and posture, regardless of its position before stimulation.[6]

Penfield's pioneering work in the 1920s and 1930s showed that the motor region of the human cerebral cortex contains a relatively accurate map of the body. As we have seen, this area is important not only for executing movements but also for planning them. The findings of the 2016 study suggest that the premotor cortex, at least of the monkey brain, also contains maps of the space surrounding the body and "action maps" of limb posture and the repertoire of actions that can be performed within the surrounding space. The initial discovery of these action maps proved to be controversial, because of the prolonged electrical stimulation involved, and because the findings challenged the traditional view of motor cortex function. Since their discovery, however, action maps have also been found in the brains of a wide variety of mammalian species other than primates, from mice and rats to squirrels and cats, using a range of methods.

Neuroscientists call our personal space the "peripersonal space," and think of it as an interface between our body and the environment, a workspace, or "action space" within which we interact with objects. Some also think of it as a "safety zone" that envelops our body. In *The Hidden Dimension,* Hall suggested that our personal distance extends to arm's length, and researchers now use this as an arbitrary measure of the peripersonal space. Most researchers agree that the brain treats this "near space" differently from space that is farther away, but its exact limits are still unclear. Some studies find marked differences in the brain's representations of space within arm's reach and the space farther away, suggesting that there is a clear boundary between near and far spaces, but others find a much smoother transition from near to far.

Nevertheless, there is a long line of evidence suggesting that peripersonal space is encoded in the right parietal lobe. In 1941, the aptly named neurologist Walter Russell Brain described two groups of patients who experienced different kinds of visual disorientation as a result of damage to the right hemisphere. In one group were three patients who had great difficulty picking up nearby objects, or pointing to farther ones, on the left side of the visual field, and in the other group were three patients who had completely lost the ability to follow familiar routes because they tended to turn left and not right. Brain attributed all of these behaviors to "an inattention to, or *neglect* of, the left half of external space," and concluded that estimations of what he called "grasping distance" and "walking distance" are dependent upon the right parietal lobe.[7]

A related phenomenon, called attentional bias, provides further insight into how we perceive near and far space. Just as the left hemisphere of the brain controls and monitors the right side of the body, and the right hemisphere controls and monitors the left, so each hemisphere also tends to direct its attention to the opposite side. This is not entirely symmetrical—the left hemisphere pays slightly more attention to the right side than the right hemisphere does to the left, resulting in an overall bias in attention to the right.

When it comes to how we perceive peripersonal space, however, this bias is reversed. In the lab, neglect is often measured with the "line bisection task." When study participants without brain damage see horizontal lines split in two by a short vertical line, they tend to perceive the left side of the line as being slightly longer than the right, but make these mistakes more often when the lines are shown in near space than when they are farther away. Furthermore, as one 2002 study showed, transcranial magnetic stimulation (TMS) of the right parietal lobe alters performance on this task, shifting attention rightward. All of this supports the idea that near space is encoded by the right parietal lobe, and that the brain makes a distinction between near and far space.[8]

Researchers have therefore used the line bisection task to investigate the boundary between near and far space. In one 2007 study, participants used a laser pointer to bisect lines presented to them at various distances both within and beyond arm's reach. Their accuracy scores showed a gradual left-to-right shift in attentional bias between the near and far distances, but also revealed consistent differences in the rate of the shift related to arm length. Those with longer arms showed a more gradual attentional shift, indicating that their near space was larger than those with shorter arms, and that the extent of near space was directly proportional to arm length.[9]

Following up on these results in 2011, the researchers used this same method to examine the size of peripersonal space in participants diagnosed with claustrophobia, or an irrational fear of enclosed spaces and crowded areas. Once again, they asked participants to bisect lines in near and far space using a laser pointer, and they used the rate of shift in attentional bias to determine the size of each participant's near space. This showed a close relationship between the size of near space and the participants' subjective reports of the amount of anxiety they feel—the greater their fear of enclosed spaces, the larger their peripersonal space. Given their 2007 finding that near space increases with arm length, the researchers also examined whether participants with longer arms experience more claustrophobia, but they found no association between the two. This may be because peripersonal space serves several different functions—arm length may affect its guidance of actions, whereas claustrophobia may impact its role in protecting the body.[10]

Our peripersonal space, then, is a multisensory area immediately surrounding our body. Our brain integrates tactile (touch) and visual stimuli that appear within it, to give us the sense that the space "belongs" to us. But sound and hearing also play a role, and studies involving a sound-touch interaction task suggest that auditory and tactile stimuli may also alter the boundaries of our peripersonal space. In one 2015 study, blindfolded participants listened to different kinds of sounds played to them on a pair of loudspeakers while touch was applied to their right middle finger. Each sound lasted for three seconds. Some sounds became more intense and were perceived by participants to be approaching their right hand, whereas other sounds became less intense and were perceived as receding from their hand; still other sounds had a constant intensity. These sound dynamics influenced the participants' perceptions of the tactile stimuli—they recognized touch applied in conjunction with approaching ("looming") sounds fractions of a second earlier than touch applied during receding sounds. Hearing the sound of a woman screaming also reduced participants' reaction times to touch applied to their finger, whereas hearing the sound of a baby laughing had the opposite effect. Thus hearing sounds that induce emotions appears to shape the boundaries of the peripersonal space, depending on the emotion the sounds induce. Hearing the sound of a baby's laughter induces positive emotions and expands the peripersonal space. By contrast, hearing the sound of a woman screaming induces negative emotions and contracts the space, as does hearing a "looming" sound, because it often signifies an approaching object, which may pose a threat.[11]

Researchers in Geneva used a variation of this approach in their 2015 study to explore the peripersonal space around different parts of the body. They found that auditory and visual stimuli enhanced the processing of touch stimuli applied to the hands, face, and trunk, but to different extents. Their results suggest that the size of the peripersonal space differs around each body part, with the near space around the torso being the largest, and that around the hand being the smallest. These experiments also showed that whereas approaching sound and visual stimuli contracted the peripersonal spaces around all three of these body parts, receding stimuli expanded only the peripersonal hand space. This is in line with the idea that peripersonal space serves a protective function, and that the peripersonal space around the hand serves specialized functions of goal-directed action, such as manipulating objects and using tools.[12]

The peripersonal space, then, is not simply an invisible bubble or zone, as Edward T. Hall envisioned in the 1960s, but appears to be composed of distinct bubbles or subzones surrounding the body parts, one for each part. It is still often thought of as a single zone with a distinct boundary, within which stimuli of different kinds are integrated to evoke enhanced responses in the brain and changes in behavior, despite conflicting results on exactly where the boundary lies, and on the nature of the transition between "near space" and "far space." The action field theory attempts to reconcile these results—as a 2018 study explains, it reconceptualizes the peripersonal space "as a set of graded fields describing behavioural relevance of actions aiming to create or avoid contact between objects and the body."[13]

Tool Use

The iconic opening sequence of Stanley Kubrick's *2001: A Space Odyssey*, depicts "the dawn of man." In prehistoric times, a band of apes, struggling for survival on a vast, barren savanna, is baffled by the sudden appearance of a large black monolith. Later on, one of them sits beside the skeleton of a larger animal, picks up a bone, and wields it as a weapon for hunting prey and fighting rivals. It then throws the bone up into the air, and as it comes spinning back down to the ground a fast cut takes us to a shot of a spacecraft orbiting the Earth. The implication is that tool use was a driving force for human evolution, which led, millions of years later, to the technology of the space age.

Indeed, it does seem that representations of tools have a special place in the brain, in which they are closely associated with those of the hand. The occipital lobe, located at the back of the brain, is largely devoted to processing visual information and so, too, are neighboring areas of the temporal lobe. Brain scanning shows that the junction of the occipital and temporal lobes contains discrete subregions that respond selectively to different categories of objects and their properties and that clearly distinguish between animate objects, such as animals and human body parts, and inanimate ones, such as tools; they encode information about what objects look like, how they move, and how they can be used. These subregions of the visual system are crucial for conceptual knowledge, as evidenced from patients with localized damage to them, damage that causes an inability to identify one or another specific category of objects.

Hands and tools look very different and belong in separate object categories. Yet there is within the left occipito-temporal junction a specialized region where the preference for hands overlaps with the preference for tools, which is activated by photographs of both hands and tools, but not by images of other body parts, other inanimate objects, or motion. We know that experience is critical for shaping the structure and function of the developing brain. Surprisingly, though, as one 2017 study has found, this overlapping hand-tool region is intact both in people who are born without hands and in those who are born blind. The hand-tool region thus appears to be innate and, developing independently of experience, to serve an important evolutionary function.[14]

A wealth of experimental data published over the past twenty-five years shows that using a tool modifies both the peripersonal space around the hand and the body schema. In an early study published in 1996, Japanese researchers trained macaques to pull distant objects toward them with a rake, while recording the activity of neurons in the primary somatosensory cortex. They found large numbers of "bimodal cells" that responded to both touch and visual stimuli, firing when the monkeys used their hand, and when the researchers applied touch to the hand or moved pieces of food toward it. As the monkeys used the rake, their visual receptive fields increased dramatically in size, responding to food not just when it was close to their hand, but also when it appeared along the entire length of the rake and in any position in space accessible with it. The cells responded in this way, although they did so only briefly, even when the monkeys were not looking at the food—the receptive fields contracted to their normal size after about three minutes, even when the monkeys only held the rake but did not use it.[15]*

Patients with brain damage provide further evidence that using a tool expands the peripersonal hand space. Some patients with damage to the right hemisphere fail to recognize touch on their left hand when they feel touch on their right hand at the same time, a phenomenon called cross-modal extinction. But when touch on the left hand is accompanied by visual stimuli in the right visual field, the severity of extinction depends on the distance between the visual stimulus and the right hand. Thus patients in a 2000 and a 2005 study almost always failed to recognize touch on the left hand when the stimuli were presented very close to their right hand, whereas the extinction was much less severe when the stimuli were presented farther away. When patients used a rake with their right hand, however, extinction in far space became severe; that is, visual stimuli presented at the end of the rake competed with touch applied to their left hand because using the rake had expanded the peripersonal space around the right hand. This depended on actually using the tool: merely holding it did not have the same effect. The extent to which tool use extended the peripersonal space is not directly related to the length of the tool, however. A rake 30 centimeters (12 inches) long caused extinction to the same distance that a rake 60 centimeters (24 inches) long did, but the effect was not as severe, and a "hybrid" rake 60 centimeters long with teeth placed halfway along its length had the same effect that a rake 30 centimeters long did. This shows that peripersonal space extends *beyond* the tip of a tool, and

* Lead researcher Atsushi Iriki reportedly conceived of this experiment in a hotel casino during a family vacation in Okinawa, while watching a croupier use a rake to move stacks of chips around the roulette table. He decided to train monkeys to use a rake for food, but "met strong criticism from people who said that Japanese macaques cannot use tools at all." He attempted it nevertheless; though at first the monkeys seemed unable to learn the task, they managed to do so after several months of training.

that the size of the extension does not depend on the length of the tool per se, but on the range within which the tool can serve its purpose.[16]

Henry Head and Gordon Holmes predicted as much over one hundred years ago. "It is to the existence of . . . 'schemata' that we owe the power of projecting our recognition of posture, movement and locality *beyond* the limits of our own bodies to the end of some instrument held in the hand," they wrote in their monumental 1911 paper. "Without them we could not probe with a stick, nor use a spoon unless our eyes were fixed upon the plate. Anything which participates in the conscious movement of our bodies is added to the model of ourselves and becomes part of these schemata: a woman's power of localization may extend to the feather in her hat."[17] Macdonald Critchley elaborated on this in his 1950 review "The Body-Image in Neurology":

> One's personality may to some extent include one's clothing. . . . A motor-car driver; a cyclist; the pilot of an aircraft—all have for the time being a body-image which includes their vehicle. Thus a motorist who worms his way in and out of city traffic is limited not by his anatomy but by the distance of his bonnet and his off-side mudguard. This type of enhanced body-image does not apply of course to the learner but only to the experienced motorist, cyclist, or pilot. Moreover, it does not apply to the passenger within the moving vehicle, for his body-image rarely attains the same measure of unity with the machine as that of the one who guides and directs it.
>
> Inanimate objects other than clothing may be taken into the body-image. Thus a blind man groping his way among unfamiliar surroundings may be regarded as possessing a body-image which extends as far as the ferrule of his stick. A surgeon working with a probe inside a patient's body-cavity, and a soldier confidently manipulating a lethal weapon, are examples of a body-image extending to include a tool. In such cases, skill is necessary, for a novice awkwardly fumbling with an instrument cannot yet be said to have incorporated the inanimate object within his body-image.[18]

Modern brain research has found Critchley's characterization of the body image to be highly accurate. In 2009, a team of researchers in Lyon, France, confirmed the hypothesis that tool use modifies the body schema by asking participants to grasp objects with a mechanical grabber 40 centimeters (16 inches) long and recording the movements of infrared emitting diodes placed on their hands and the grabber with a high-resolution 3-D motion-tracking system. Using the grabber not only altered the grasping movements they later made without it, but also affected touch perception on the arm they had used the grabber with—they took longer to initiate grasping movements, their movements were slower, though they remained accurate, and they perceived touches to the elbow, wrist, and fingertip to be farther apart than they actually were. Using the grabber had lengthened the brain's representation of the arm, a change that persisted for up to fifteen minutes. This shows that the body schema is highly dynamic and can be "updated" quickly to incorporate tools.[19]

Almost a decade later, in a 2018 study, when participants held a wooden stick in one hand, they could sense touch along its entire length. When they moved the stick themselves to tap it against another object, they could localize the touch with high accuracy, and when the stick was moved for them, their estimates were only slightly less accurate. The small differences probably occurred because actively moving the stick provides more feedback—motor signals on top of the vibrations produced when the stick comes in contact with another object. The nervous system therefore treats the stick as an extension of the body, so that the sense of touch is extended along its entire length.[20]

Using electroencephalography (EEG), in a 2019 follow-up study, the researchers went on to show that the brain responds to touch on the stick in much the same way that it does to touch on the body. Applying touch to a stick that participants held in their right hand almost immediately activated the left primary somatosensory cortex, and the activity then quickly spread back to the parietal lobe. This same pattern of activity was seen in a cancer survivor who had lost her sense of the position and movement of her right arm (proprioception) following removal of a brain stem tumor, supporting the researchers' earlier conclusion that the ability to localize touch on the stick depends on vibrations. The brain therefore co-opts the mechanisms for sensing touch on the body to sense touch applied to "embodied" tools (those held or used by the body).[21]

A 2009 magnetic resonance imaging (MRI) study performed on monkeys further shows that tool use is associated with long-lasting changes in the brain's gray and white matter. After two weeks of intensive training in how to use a rake, three macaques exhibited increases in gray matter volume within the somatosensory, parietal, and temporal regions associated with tool use, as well as an increase in white matter volume in the cerebellum.[22]

The brain thus incorporates a hand-held tool into its representations of the body and surrounding space. Many studies have concluded that this requires active use of the tool, but there is also evidence to suggest that active use is not always necessary. In one 2017 study, participants saw a mirror reflection of their right hand using a mechanical grabber, giving them the impression that their left hand was using the grabber. Afterward, they perceived pairs of touch stimuli on their right arm to be farther apart than they actually were, indicating that use of the grabber had lengthened the neural representation of the arm. But perception of touch on their left arm was altered in exactly the same way, even though it had been under the table and out of sight all along.[23] Another study, in 2014, showed that participants' merely thinking about using a tool altered the kinematics of their subsequent movements, suggesting that mental imagery is sufficient to incorporate the imaginary tool into the body schema.[24]

Tool use is thus a window into the plasticity of body representations. The experiments performed on monkeys, on patients with brain damage, and on controls show that using a tool not only reorganizes the brain's representations of the body but also expands the peripersonal space. Thus, as the 2014 study clearly shows, the brain treats the tool not as an external object, but as an extension of the body, endowed with sensing capabilities. The tool becomes "embodied," in effect, and this facilitates gaining mastery of its use.[25]

The peripersonal space, whose properties appear to be very similar to those traditionally attributed to the body schema, may turn out to be an external component of the body schema—or perhaps the terms "body schema" and "peripersonal space" actually refer to different aspects of the same representation. Some researchers use the two terms interchangeably, and the Lyon team suggests that they may "be considered as the two faces of the same coin."[26]

Prosthetic Limbs

In 2000, German archaeologists excavating at the necropolis of Sheik Alb-el-Qurna in West Thebes, near modern-day Luxor, Egypt, entered a burial chamber built by a high royal official during the Eighteenth Dynasty (ca. 1550–1300 BCE) for members of his family. As was customary in ancient Egypt, this tomb had subsequently been extended to include additional burial chambers, all connected to the burial chapel by a series of narrow shafts. One of the adjoining rooms contained the mummified remains of several individuals, along with funerary pottery typical of the Twenty-First and Twenty-Second Dynasties (ca. 1065–740 BCE). One of the mummies had been broken into several pieces, probably when the tomb was plundered by grave robbers, although it was otherwise well preserved. Fastened to its right foot with linen strapping was an artificial big toe, which is now thought to be the world's oldest prosthetic device.

Examination of the broken mummified remains indicated that they belonged to a woman, fifty to fifty-five years old, later identified as Tabaketenmut, a priest's daughter who lived sometime between 950 and 710 BCE. Scholars think she may have developed gangrene in her toe as a result of diabetes; but regardless of the cause, her toe was amputated some time before her death because the wound had healed fully, without requiring stitches, with the stump covered over by an intact layer of skin. Her artificial big toe, now on display at the National Museum of Egyptian Civilization in Cairo, is exquisitely formed and anatomically accurate, including a sculpted nail. It consists of three pieces—a long piece of dark, dense hardwood, which replaced the amputated toe, a smaller wooden plate, to which the wooden toe was attached by lace running through a series of holes, and a soft upper section made of a leatherlike material. Its maker clearly appreciated the function of the big toe as well as its structure: the device is tied snugly to the foot, its lower surface is flat, to improve stability, and its design even incorporates a rudimentary joint, to give its owner some degree of flexibility.[27]*

Improvements on this design did not appear until some 2300 years later. The German imperial knight Gottfried "Götz" von Berlichingen had two prostheses made from cast iron after losing his right hand during the siege of Landshut in 1504, the second of which had movable fingers capable of grasping objects. Toward the end of the sixteenth century, the French military surgeon Ambroise Paré designed and attempted to build prostheses that worked liked biological limbs. He designed artificial legs with mechanical knees that could lock when their user stood still, but his pièce de résistance was a mechanical artificial hand fitted with dozens of catches, springs, and sockets that allowed for each of the fingers to move independently. Dubbed

* In 2012, Jacqueline Louise Finch and her colleagues at the University of Manchester built replicas of the device and tested their functionality on two big toe amputee volunteers, who said they were "by far the more comfortable and convincing" prostheses they had ever used.

"Le Petit Lorrain," the hand was worn in battle by a French army captain, who claimed that it worked so well that he was able to grip the reins of his horse with it.[28]

The start of the twenty-first century saw the development of advanced brain-computer interfaces, which allow for prosthetic limbs controlled by the power of thought. These devices are built around microelectrode arrays that are implanted directly into the motor cortex, which read the brain activity associated with the initiation and control of movement. This information is sent it to a computer chip which "decodes" the pattern of activity and translates it into commands that can be sent to the robotic limb. An early example of this is the BrainGate neural interface system, developed by researchers at Brown University in Providence, Rhode Island, in collaboration with the medical device company Cyberkinetics, Inc. The BrainGate was first demonstrated in 2002, when the Brown University researchers implanted the device into the brain of a paralyzed volunteer, enabling him to remotely control, via the internet, a prosthetic hand over 3,400 miles away at the University of Reading, England, and then to control the movements of an electric wheelchair.[29]

By today's standards, however, even the BrainGate system seems primitive. Advances in our understanding of the neurobiological mechanisms underlying bodily awareness, combined with developments in robotics, engineering, and materials science, are aiding the development of sophisticated prosthetic limbs that not only look and feel more realistic, but that will also employ the principles of multisensory integration to become fully incorporated into the brain's representations of the body. This next generation of prostheses will include artificial skin and nerves equipped with miniaturized synthetic receptors that send multisensory feedback to the brain. It may sound like science fiction, but devices with these properties and capabilities are already under development.

In 2018, researchers at University College London published one of a very few studies of neural representations of artificial limbs in amputees. Working in collaboration with colleagues in the Netherlands, England, Canada, and Israel, they recruited thirty-two one-handed and twenty-four two-handed people as participants, matched for age and sex, and used a questionnaire to find out about their daily prosthesis usage habits. Half of the one-handed participants had had one of their hands amputated, and the other half had been born missing one hand. Two of them did not own an artificial limb, and another three had one but weren't using it at the time of the study. The rest used an "active prosthetic limb," or a cosmetic one, or both, regularly.

All of the participants then underwent functional neuroimaging while they viewed images of objects, hands, and different kinds of prostheses, and this revealed a relationship between activity in the hand-selective region at the occipito-parietal junction and usage of a prosthetic limb, with significant differences between the two groups of participants. Images of artificial limbs evoked a far stronger response in the one-handed than in the two-handed participants, and the response was stronger in the one-handed participants who used a prosthesis regularly than in those who did not. What's more, the responses in the one-handed participants were not specific to the types of prostheses shown in the images, or to their familiarity with them—images of unfamiliar prostheses evoked responses of the same strength as images of their own. Those who used their artificial limb regularly also exhibited stronger functional connectivity between this hand-selective brain area and the somatosensory and motor areas devoted to the hands. The researchers conclude that artificial limbs activate brain areas involved in encoding body representations and that using a prosthesis leads to embodiment of the device. Importantly, the results also suggest that the more an amputee uses a prosthesis, the greater the extent of its embodiment.[30]

Some amputees do not use an artificial limb regularly, however, and some do not use one at all. Prostheses do not feel like real limbs, and users do not have full control over them, so the challenge is to build devices that provide sensory feedback to the brain. To this end, physicians and engineers at the Rehabilitation Institute of Chicago and Northwestern University developed a novel surgical procedure called targeted muscle reinnervation, whereby residual nerves in the stump are rerouted to chest muscles that previously had been connected to the missing limb. With their new nerve supply, these muscles serve to amplify the movement commands of the amputated nerve, and can be used to control a robotic arm. When done successfully, this procedure does indeed improve control over a prosthesis. Normally, the motor commands for clenching the fist are conveyed to the hand via the median nerve. If, after amputation, the residual median nerve is surgically transferred to the chest (pectoralis) muscle, that muscle can then generate the hand-close signal. When the amputees think about clenching their fist, the reinnervated pectoralis muscle will contract, and the electrical signal will command the motorized hand to close.

Targeted muscle reinnervation can restore at least some sensory functions, too. In 2006, former marine Claudia Mitchell, who lost her arm in a motorcycle accident at the age of twenty-four, became the world's first female recipient of a "bionic" arm controlled in this way. Three months after her surgery, she started to feel tingling sensations in her missing hand when her chest was touched. By five months, she perceived any touch on her chest to come from her missing hand, and at six months, "she developed a relatively faint percept of her middle finger on the lateral chest wall."[31]

Researchers at the Rehabilitation Institute of Chicago have since been developing a system that delivers an artificial sense of touch to prosthetic limbs. As reported in their 2011 study, the system consists of a miniaturized robot equipped with pressure, vibration, force, and temperature sensors, which can be attached to the prosthesis with adhesive pads, and is connected to a small plunger on a plastic cuff worn around the stump. When pressure is applied to the robotic sensors, it signals the plunger to push down on the reinnervated nerve, evoking touch sensations that are projected to the missing limb, and are perceived to originate in the prosthesis. The researchers used this system to elicit the rubber hand illusion in two arm amputees—synchronous stroking of the rubber hand and the robotic sensors evoked the sensation of touch in the fake hand, suggesting that the artificial touch sensations helped the amputees to embody the fake hand.[32]

More recently, in 2018, researchers at the Massachusetts Institute of Technology in Cambridge, Massachusetts, and at Brigham and Women's Hospital in Boston constructed a muscle-nerve interface that conveys proprioceptive inputs to the brain. This consists of two muscle tendons that are surgically removed during amputation and reconnected so that one muscle stretches when the other contracts, and vice versa, when grafted into the amputees' residual limb. Having performed some proof-of-concept studies in mice, the researchers constructed two of these "agonist-antagonist myoneural interfaces" (AMIs, pronounced "AY-mees") and grafted them into the residual limb of a fifty-three-year-old below-the-knee male amputee before he was fitted with a prosthetic ankle and foot.* The transplants contain muscle spindles and mechanoreceptors, so that when the amputee moves his prosthesis, these receptors transmit information to the residual nerves in the stump, which then carry them up through the spinal cord to the brain. The brain interprets these signals as sensations of joint position and speed, enabling it to relate motor commands to errors in movement, which improves control of the prosthesis.[33]

There is a far simpler method of delivering proprioceptive feedback to prosthetic limbs. In two-handed people, vibrating the elbow tendons at certain frequencies produces illusions of movement, so that the arm is perceived to have bent back into an anatomically impossible position. Amputees who have undergone targeted reinnervation also experience this "muscle vibration illusion." When vibrations are applied to their reinnervated muscles, arm amputees perceive complex movements of their missing hand, wrist, or elbow, depending on which nerve supplied the muscle being vibrated. In another 2018 study involving two upper-arm amputees, vibrating the muscles reconnected to the median nerve induced various perceptions that the fingers were flexing; vibrating muscles reconnected to the radial nerve produced percepts of finger extension; and vibration of residual biceps and triceps muscles variously produced illusory gripping, pinching, hand-opening, elbow extension, and arm rotation movements. This gave the amputees a greater sense of agency and ownership over their artificial limbs, and improved their control of the devices, although it also induced phantom sensations that interfered with the feedback.[34]

State-of-the-art prostheses can deliver multisensory feedback to their users. One such device, developed by researchers in Switzerland and Italy, employs a robotic hand in a hybrid approach to simultaneously deliver touch and proprioceptive feedback. The robotic hand has sensors integrated

* Hugh Herr, the principal investigator on this project, is himself a double-lower-leg amputee. As a rock climber prodigy, he scaled Mount Temple in the Canadian Rockies when he was only eight. While climbing Mount Washington in 1982, Herr was caught in a blizzard, and spent three nights in temperatures of –20°F. This caused severe frostbite, and he had to have both of his legs amputated below the knee. Undeterred, Herr began climbing again after months of surgery and rehabilitation, using specialized prostheses he had designed himself. He went on to earn a master's degree in mechanical engineering from MIT and a PhD in biophysics from Harvard; he has spent his academic career developing artificial limbs and wearable augmentation robotics and now holds more than 100 patents for such devices.

into each of its digits, which send position and force information to a computer. The information is translated into trains of electrical impulses that are then sent to multichannel electrodes implanted directly into the ulnar and median nerves in the upper arm. With a little training, two lower-limb amputees fitted with this robotic hand could detect passive movements of the robotic hand, identify the size and shape of objects they grasped with it, and discriminate foam from plastic from their consistency.[35]

Sensory feedback confers other advantages. A prosthetic leg equipped with foot and knee sensors plugged directly into the residual tibial nerve of the stump gave amputees the sensations both of knee movement and of the sole of their missing foot touching the ground. They walked faster and took longer to fatigue with the sensory feedback than without it. (The finding that sensory feedback reduced oxygen consumption during walking lends credence to the argument that advanced prostheses give paralympic athletes an unfair advantage.) In a 2019 study, researchers found that feedback also reduced the phantom pain sensations perceived to originate in their missing limb.[36] Another 2019 study provided some evidence that sensory feedback can also reduce distorted perception of a phantom limb, such as telescoping—the perception that a phantom limb is shorter than the missing limb.[37]

The prostheses described above will soon be superseded by even more advanced devices. Thus researchers at the BioRobotics Institute in Pisa, Italy, have developed an artificial fingertip equipped with an array of neuromorphic sensors that can convert touch information into electrical impulses; biomedical engineers at Johns Hopkins School of Medicine have invented a multilayered electronic skin that enabled an amputee to perceive touch and pain sensations from a prosthetic hand; engineers at Cornell University have developed a prosthetic hand containing optical strain sensors, capable of picking out a ripe tomato by its softness; materials scientists and engineers at the National University of Singapore have developed a prototype of an electronic skin that can transmit touch and temperature information via thousands of artificial receptors; and researchers at the École polytechnique fédérale de Lausanne in Switzerland have just prototyped soft and stretchable pneumatic skin that provides touch feedback via a thin metal film of strain sensors.[38]

Artificial limbs can also be permanently anchored to bone at the amputation stump with screws made from rustproof metals, such as titanium, or their alloys. New bone cells attach themselves to the fixture by a process called osseo-integration, keeping the prosthesis firmly attached in place for long periods of time.[39] Indeed, neuroprosthetic technology is advancing so rapidly that users would likely have the option to upgrade their devices regularly. Cost is, of course, a major obstacle to obtaining a state-of-the-art prosthesis, but technologies such as 3-D printing are already making artificial limbs more affordable and available to more people.

10 Conclusion

> The body is a big sagacity, a plurality with one sense, a war and a peace, a flock and a shepherd. . . . Behind thy thoughts and feelings, my brother, there is a mighty Lord, an unknown sage—it is called Self; it dwelleth in thy body, it is thy body.
>
> —Nietzsche, *Thus Spoke Zarathustra*

The mind and body have always been regarded as distinct from each other. This "dualism," usually attributed to the French mathematician and philosopher René Descartes (1596–1650), separates the mental realm of the mind from the physical realm of the brain and body. Most modern neuroscientists reject Cartesian mind-body dualism, however, and consider the mind to be a product of the brain. Even so, the ghost of dualism still haunts science and medicine, and we continue to draw a distinction between mind and body. This is evident from terms such as "psychosomatic" (from the Greek *psyche,* meaning "mind" and *soma,* meaning "body"), which is used by physicians to denote a physical illness caused, or at least exacerbated by, some mental factor, the implication being that there is some kind of interaction between the mental and the physical.

In the twenty-first century, popular thinking about the mind and brain has gone from one extreme to the other. The idea that the mind has a neurological basis has taken root. Disciplines such as psychology and psychiatry, which traditionally dealt with "mental" aspects of the human condition, are increasingly drawing on advances in neuroscience, and this combined approach is proving to be fruitful. At the same time, interest in neuroscience has grown enormously, and the idea that we are our brains is now firmly established in the collective mind. This "neurocentric" view is ill conceived, however, and creates its own form of modern dualism, which regards the brain and body as separate entities. But the brain does not exist in isolation; it is one part of a complex and dynamic system that also includes the body and the environment, and we are the product of ongoing interactions between these nodes. Within this system, information flows in all directions. We perceive our body within the world, and act upon our perceptions, but our actions also influence the way we perceive things. Although our body lies at the intersection of all of these processes, it is far more than a passive conduit for our actions and perceptions.

The body is both an object and the subject. We do not perceive the world as it really is; rather, our perception of the world is our brain's best guess at reality, a neural construct built from the limited information it receives through our senses. This is also true of our body. To a large extent, we perceive our body in the same way that we perceive an object in the outside world, through multiple channels of sensory information that enter our brain: the sight of our body as it moves, the sounds it makes, the touch and pain signals that arise from our skin, our muscle sense, and our internal sensations. Our brain integrates all of this information to generate models of our body and the space surrounding it. Our body is our brain's user interface, an object—or tool—with which we perceive and interact with the world. But it is an object unlike any other. We do not perceive our body objectively; we perceive it subjectively, from the inside, and the end result of our bodily perception is what each of us calls "me." As well as enabling us to act, our body plays an active role in how we perceive the world and makes itself felt in our thoughts and feelings. That is to say, our body is vital for our self-consciousness: our awareness of our self and of our place within the world is inextricably linked to our body and the sensations that arise from within and without it.

Transhumanists believe that technological advances in our ability to map synaptic connections in the brain, combined with ever-accelerating computing power, will eventually enable "mind uploading." It is only a matter of time before technology allows for high-resolution mapping of all the connections within the human brain, and supercomputers become powerful enough to store all this information. At this point, or so they assert, it will be possible to install the mind's architecture onto a computer, whereupon it will gain consciousness and we will achieve immortality.

The new science of self-consciousness certainly will allow humans to merge with machines in increasingly sophisticated ways. It is, for example, already driving the development of next-generation artificial limbs that can integrate into the user's body image, and of fully immersive virtual reality interfaces that transfer sensations to the user, among other things. At least for the more levelheaded practitioners of this science, however, it is highly unlikely that mind uploading will ever be anything more than a fantasy. Eventually, it probably will be possible to upload the neural architecture of the mind to a supercomputer—but whether the upload can reconstitute an individual's consciousness is another matter altogether. Crucially, the uploaded "mind" would lack a body and therefore would be unable perceive the world, act upon it, or gain self-awareness. For this same reason, it is highly unlikely that lab-grown "minibrains" could ever become conscious either.[1]

On the other hand, this new understanding of self-consciousness is forcing us to redefine what it means to be conscious, and which organisms possess consciousness. For untold ages, religious thinkers, philosophers, and naturalists alike placed humans at the pinnacle of life on Earth and denied other animals the possibility of consciousness. In recent years, however, this view has begun to change in light of growing evidence that cognitive capabilities once thought to be unique to humans are in fact present in an ever-wider variety of other animals. For example, scrub jays store food in various locations as winter approaches, but their food stores can be pilfered by other jays. They have a strategy to prevent this, however: when one jay sees that another is watching it cache its food, it waits until the potential thief has left and then recaches the food in a different location. Scrub jays can also plan for the future: in the lab, they preferentially store food in locations where they have learned they will be hungry the following morning. Similarly, on the Southwest Pacific island of New Caledonia, the native crows can keep in mind the function and location of three different tools, then use them to perform complex sequences of actions by planning several moves ahead.[2] Dolphins, whales, elephants, and various species of monkeys can recognize themselves in a mirror—a yardstick for self-awareness—and the list of animal species that can pass this mirror-self-recognition test continues to grow.

Furthermore, other kinds of animals are also aware of their bodies, at least to some extent. The hermit crab has a soft abdomen and inhabits empty seashells, which protect its fragile exoskeleton. As it increases in size, it outgrows and regularly abandons its shell for another, slightly larger one. But a hermit crab never moves into a shell that is too large for its body. Instead, when it finds an oversized shell, it waits nearby for other hermit crabs to see it. Soon enough, a group of them forms a queue around the empty shell, and when one of them determines it's the right size and occupies it, another hermit crab, perhaps the first waiting one, abandons its shell for the new occupant's old shell—provided, of course, the old shell is the right size. Before long, shell swapping is taking place en masse, with each hermit crab finding the best-fitting new home. The hermit crab favors the shells of sea snails because its abdomen is adapted to latch onto the insides of such shells. Competition can be fierce, however, and the crabs have been known to use another mollusk shell, or even a tin can, as an alternative when a sea snail shell is not available.

Thus the hermit crab has a strong sense of the size and shape of its body and accurately perceives changes in its bodily dimensions in order to find a new shell of the appropriate size. Hermit crabs can adapt to bodily changes in other ways. Some species attach large sea anemones to one side of their sea snail shell to counterbalance the shell's coiled spiral and smaller anemones near the shell's opening to protect against octopus attacks. In one set of experiments, researchers placed twelve terrestrial hermit crabs into a corridor with alternating left and right corners. When the crabs negotiated a tight corner, they rotated their body to avoid coming into contact with the wall, and when the researchers attached a plastic plate to their shell, the crabs increased the angle of their body rotation, as if they had assimilated the plastic plate into their body. It is thus tempting to conclude that the terrestrial hermit crab has conscious awareness of a dynamically updated body image; at the very least, we can say that it must possess some kind of bodily self-consciousness.[3]

Insects also appear to possess bodily awareness to some extent. Bumblebees, for example, vary widely in size, live in nests that sometimes contain hundreds of individuals, and often negotiate dense, cluttered vegetation while foraging for food. In 2020, researchers in Germany trained bumblebees, whose wingspan is much wider than their body is long, to fly through a tunnel with walls containing gaps of different sizes, and found that they could fly smoothly through those which were much narrower than their wingspan. As they approached each gap, the bees performed small side-to-side maneuvers to scan the size of the opening, then rotated their body to reduce their frontal width and avoid collision. The smaller the gap, the longer the bees spent assessing its size; at the narrowest gaps, they reoriented their in-flight posture by 90 degrees and flew through sideways.[4]

Even fruit flies (*Drosophila melanogaster*), which are just 3 millimeters long with a pinhead-sized brain containing just 100,000 neurons, have a basic sense of bodily awareness. They learn their body size through motion parallax—the apparent velocity of stationary objects relative to the observer's movement, by which closer objects appear to move faster. In their 2019 study, researchers identified a small group of *D. melanogaster* brain cells that encode a long-term memory of body size and showed that interfering with these cells erases the memory.[5]

If bodily awareness is fundamental to self-consciousness, then it follows that hermit crabs, bumblebees, and fruit flies must possess at least some degree of consciousness. Bodily awareness can also be programmed into machines. In 2006, engineers at Cornell University described a four-legged "Starfish" robot, with a built-in body plan that automatically updates itself. If one of its legs is damaged, or removed altogether, the machine detects the change and compensates for the damage by adjusting its gait. Newer iterations have machine learning algorithms that allow them to model themselves without any prior knowledge of their own shape or the laws of physics.[6] These self-modeling capabilities closely resemble bodily awareness. Do they constitute self-consciousness?

Consciousness is the biggest scientific mystery of all. The past twenty years have been marked by major progress in our understanding of bodily awareness, and this has changed the way we think about the nature of consciousness. But scientific discovery always raises more questions than it answers, and, despite these remarkable advances, we are still a very long way from understanding consciousness or solving the puzzle of the self. We may never fully understand our selves, but our knowledge and technology will surely continue to move forward and enable us to repair our damaged bodies and to modify and improve our selves in ever more sophisticated ways.

Acknowledgments

I am deeply grateful to Bob Prior, Matt Browne, and Chris Eyer at the MIT Press for their support and encouragement, and also for their patience with this project, which took far longer to complete than was expected. I would also like to thank the researchers, technicians, and other lab members for taking the time and trouble to discuss and demonstrate their work: Chris Berger, Olaf Blanke, Claudio Brozzoli, Lucilla Cardinali, Laura Case, Henrik Ehrsson, Alessandro Farné, Arvid Guterstam, Matt Longo, Lara Maister, Tamar Makin, Robin Mange, Paul McGeoch, Luke Miller, Hyeong-Dong Park, Vilayanur Ramachandran, Alice Roy, and Manos Tsakiris. Finally, I thank Holly for her love and support, her kindness and patience, and her help with the manuscript.

Glossary

agency, sense of — the sense that we are in control of our body, our thoughts, and our actions; a core component of **bodily awareness**

agnosia — the inability to recognize or identify faces or objects

Alice in Wonderland Syndrome — a rare neurological condition characterized by hallucinations in which visual perception of the body and objects is distorted

alien hand syndrome — a rare neurological condition in which a limb seems to act on its own free will (also known as "Dr. Strangelove syndrome" or "anarchic hand sign")

anorexia nervosa — an eating disorder characterized by weight loss, difficulty maintaining an appropriate body weight, distorted **body image** and body dissatisfaction

autotopagnosia — a form of **agnosia** characterized by the inability to localize, recognize, or identify parts of one's own body

bodily awareness — a basic form of self-consciousness consisting of the senses of **agency** and **body ownership**

body image	the perception of, and attitudes toward, one's own body, as defined by Paul Ferdinand Schilder
body integrity identity disorder (BIID)	a rare neurological disorder in which one desires to amputate a healthy limb
body ownership, sense of	the sense that our body belongs to our self; a core component of **bodily awareness**
body schema	a postural model of the body, as defined by Henry Head and Gordon Holmes
cerebral cortex	the thin outermost portion of the mammalian brain
corpus callosum	a massive bundle of nerve fibers connecting the two hemispheres of the brain
electroencephalography (EEG)	a neuroimaging technique that records the electrical activity of large populations of neurons in the brain through scalp electrodes
frontal lobe	one of four major lobes of the brain, involved in higher cognitive functions such as attention and problem solving, and containing the **motor cortex**
functional magnetic resonance imaging (fMRI)	a neuroimaging technique that measures changes in cerebral blood flow as a proxy for brain activity levels
insula	a small region of the **cerebral cortex**, located deep within the fissure separating the frontal and parietal lobes, known to play an important role in self awareness

magnetoencephalography (MEG) a neuroimaging technique that maps brain activity by recording the magnetic fields generated by the electrical activity of populations of nerve cells

motor cortex a region of the cerebral cortex, located at the back of the **frontal lobe** involved in the planning, control, and execution of voluntary movements

multisensory integration the process by which information from different sensory modalities (such as sight, touch, hearing, and **proprioception**) are combined by the brain

nociceptor a small diameter primary sensory neuron that detects painful stimuli

parietal lobe one of the four major lobes of the brain, involved in **multisensory integration**, and containing the **somatosensory cortex**

phantom limb the sensation that an amputated or missing limb is still attached to the body

proprioception our sense of the position and movement of our body and of our limbs, joints, and other body parts in space, sometimes referred to as "muscle sense"

rubber hand illusion an illusion of **bodily awareness**, in which one takes ownership of a fake hand

somato- a prefix or combining form, meaning "related to the body"

somatoparaphrenia a delusion of disownership of body parts

somatosensory cortex a region of the **cerebral cortex** that processes touch information from the body, located at the front of the **parietal lobe**

somatotopy — the point-to-point correspondence of a body part and its neural representation, such that adjacent body parts are "mapped" to adjacent areas of the brain

synesthesia — a neurological condition in which stimulation in one sensory pathway elicits simultaneous sensations in another

Notes

Chapter 1

1. Holmes, R. (2016), *The Hottentot Venus: The Life and Death of Saartjie Baartman, Born 1789—Buried 2002*, London: Bloomsbury; Qureshi, S. (2004), Displaying Sara Baartman, the "Hottentot Venus," *History of Science*, 42:233–257, 236.

2. American Society of Plastic Surgeons (2019), *2019 National Plastic Surgery Statistics*. Available at https://www.plasticsurgery.org/documents/News/Statistics/2019/plastic-surgery-statistics-report-2019.pdf.

3. Furnham, A., and Levitas, J. (2012), Factors that motivate people to undergo cosmetic surgery, *Canadian Journal of Plastic Surgery*, 20:e47–e50.

4. Renee Engeln, as quoted in Silver, K. (2017), When obsession with beauty becomes a disease, *Pacific Standard*, April 18.

5. Koch, C., Massimini, M., Boly, M., and Tononi, G. (2016), Neural correlates of consciousness: Progress and problems, *Nature Reviews Neuroscience*, 17:307–321.

6. Favazza, A. R. (1996), *Bodies under Siege: Self-mutilation and Body Modification in Culture and Psychiatry*, 2nd ed., Baltimore: Johns Hopkins University Press.

7. James, W. (1918), *The Principles of Psychology*, vol. 1, New York: Henry Holt, p. 242.

Chapter 2

1. Mitchell, S. W. (1866), The case of George Dedlow, *Atlantic Monthly*, July, 1–11; Kline, D. G. (2016), Silas Weir Mitchell and "The Strange Case of George Dedlow," *Neurosurgical Focus*, 41:E5, DOI: 10.3171/2016.4.FOCUS1573.

2. Bourke, J. (2009), The art of medicine: Silas Weir Mitchell's The Case of George Dedlow, *Lancet*, 373:1332–1333; Figg, L, and Farrell-Beck, J. (1993), Amputations in the Civil War: Physical and social dimensions, *Journal of the History of Medicine and Allied Sciences*, 48:454–475.

3. Finger, S. (1994), *Origins of Neuroscience*, Oxford: Oxford University Press.

4. Bailey, P. (1958), *Silas Weir Mitchell, 1829–1914*, Washington, DC: National Academy of Sciences.

5. Mitchell, S. W. (1872), *Injuries of Nerves and Their Consequences*, Philadelphia: J. P. Lippincott.

6. Berger, I. H., and Bacon, D. R. (2009), Historical notes on amputation and phantom limb pain: "All Quiet on the Western Front?," *Gundersen Lutheran Medical Journal*, 6:26–29; Drucker, C. B. (2008), Ambroise Paré and the birth of the gentle art of surgery, *Yale Journal of Biological Medicine*, 81:199–202; Ambroise Paré, as quoted in Wade, N. J. (2003), The legacy of phantom limbs, *Perception*, 32:517.

7. René Descartes, as quoted in Wade, N. J. (2003), The legacy of phantom limbs, *Perception*, 32:518.

8. Kline, D. G. (2016), Silas Weir Mitchell and "The Strange Case of George Dedlow," *Neurosurgical Focus*, 41:E5, DOI: 10.3171/2016.4.FOCUS1573.

9. Bourke, J. (2009), The art of medicine: Silas Weir Mitchell's The Case of George Dedlow, *Lancet*, 373:1332–1333.

10. Riddoch, G. (1941), Phantom limbs and body shape, *Brain*, 64:197–222.

11. Kline, D. G. (2016), Silas Weir Mitchell and "The Strange Case of George Dedlow," *Neurosurgical Focus*, 41:E5, DOI: 10.3171/2016.4.FOCUS1573.

12. Bourke, J. (2009), The art of medicine: Silas Weir Mitchell's The Case of George Dedlow, *Lancet*, 373:1332–1333.

13. Riddoch, G. (1941), Phantom limbs and body shape, *Brain*, 64:197–222.

14. Henderson, W. R., and Smyth, G. E. (1948), Phantom limbs, *Journal of Neurology, Neurosurgery and Psychiatry*, 11:88–112.

15. Riddoch, G. (1941), Phantom limbs and body shape, *Brain*, 64:197–222.

16. Henderson, W. R., and Smyth, G. E. (1948), Phantom limbs, *Journal of Neurology, Neurosurgery and Psychiatry*, 11:88–112.

17. Flor, H. (2002), Phantom-limb pain: characteristics, causes, and treatment, *Lancet Neurology*, 1:182–189.

18. Solonen, K. A. (1962), The phantom phenomenon in amputated Finnish war veterans, *Acta Orthoppaedica Scandinavica*, 54:1–37; Schott, G. D. (2014), Revealing the invisible: The paradox of picturing a phantom limb, *Brain*, 137:960–969.

19. Henderson, W. R., and Smyth, G. E. (1948), Phantom limbs, *Journal of Neurology, Neurosurgery and Psychiatry*, 11:88–112.

20. White, J. C., and Sweet, W. H. (1969), *Pain and the Neurosurgeon,* Springfield, IL: C. C. Thomas, cited in MacLachlan, M. (2004), *Embodiment: Critical, Clinical and Cultural Perspectives,* Milton Keynes, UK: Open University Press.

21. Henderson, W. R., and Smyth, G. E. (1948), Phantom limbs, *Journal of Neurology, Neurosurgery and Psychiatry,* 11:88–112; Flor, H. (2002), Phantom-limb pain: Characteristics, causes, and treatment, *Lancet Neurology,* 1:182–189.

22. Mitchell, S. W. (1872), *Injuries of Nerves and Their Consequences,* Philadelphia: J. P. Lippincott.

23. Ackerly, W., Lhamon, W., and Fitts, W. T. (1955), Phantom breast, *Journal of Nervous & Mental Disorders,* 121:177–178.

24. Dijkstra, P. U., Rietman, J. S., and Geertzen, J. H. B. (2007), Phantom breast sensations and phantom breast pain: A 2-year prospective study and a methodological analysis of literature, *European Journal of Pain,* 11:99–108.

25. Wade, N. J., and Finger, S. (2010), Phantom penis: Historical dimensions, *Journal of the History of Neuroscience,* 19:299–312.

26. Fisher, C. M. (1999), Phantom erection after amputation of penis: Case description and review of the relevant literature on phantoms, *Canadian Journal of Neurological Science,* 26:53–56.

27. Ramachandran, V. S., and Hirstein, W. (1998), The perception of phantom limbs: The D. O. Hebb lecture, *Brain,* 121:1603–1630.

28. James, W. (1918), *The Principles of Psychology,* vol. 1, New York: Henry Holt, pp. 291–296.

29. Figg, L., and Farrell-Beck, J. (1993), Amputations in the Civil War: Physical and social dimensions, *Journal of the History of Medicine and Allied Sciences,* 48:454–475.

30. Krueger, C. A., Wenke, J. C., and Ficke, J. R. (2012), Ten years at war: Comprehensive analysis of amputation trends, *Journal of Trauma and Acute Care Surgery,* 73:S438–S444.

Chapter 3

1. ABC (Australian Broadcasting Corporation) Television (2009), *Catalyst,* Body identity, May 21, Transcript available at https://www.abc.net.au/catalyst/body-identity/11011208.

2. Jean-Joseph Sue, as quoted in Johnston, J., and Elliott, C. (2002), Healthy limb amputation: Ethical and legal aspects, *Clinical Medicine,* 2:431–435.

3. von Krafft-Ebing, R. (1886/1907), *Psychopathia Sexualis. Antipathic Sexual Instinct: A Medico-Forensic Study,* New York: Rebman, pp. 235–237.

4. Jensen, R., and Jensen, R. (1975), *Amputee Love*, no. 1, Berkeley, CA: Last Gasp.

5. Money, J., Jobaris, R., and Furth, G. (1977), Apotemnophilia: Two cases of self-demand amputation as a paraphilia, *Journal of Sex Research*, 13:115–125.

6. Gilbert, M., dir. (2004), *Whole*, Sundance TV, http://frozenfeetfilm.com/whole/.

7. First, M. B. (2005), Desire for amputation of a limb: Paraphilia, psychosis, or a new type of identity disorder, *Psychological Medicine*, 35:919–928.

8. Dieguez, S., and Annoni, J.-M. (2012), Asomatognosia: Disorders of the bodily self, in Godofroy, O., ed., *The Behavioral and Cognitive Neurology of Stroke*, Cambridge: Cambridge University Press.

9. Gerstmann, J. (1942), Problem of imperception of disease and of impaired body territories with organic lesions, *Archives of Neurology & Psychiatry*, 48:890–913.

10. Botvinick, M., and Cohen, J. (1998), Rubber hands "feel" touch that eyes see, *Nature*, 391:756.

11. Moseley, G. M., Olthof, N., Venema, A., Don, S., Wijers, M., Gallace, A., and Spence, C. (2008), Psychologically induced cooling of a specific body part caused by the illusory ownership of an artificial counterpart, *Proceedings of the National Academy of Sciences*, 105:13169–13173; Barnsley, N., McAuley, J. H., Mohan, R., Dey, A., Thomas, P., and Moseley, G. M. (2011), The rubber hand illusion increases histamine reactivity in the real arm, *Current Biology*, 21:R945.

12. Siedlecka, M., Klimza, A., Łukowska, M., and Wierzchoń, M. (2014), Rubber hand illusion reduces discomfort caused by cold stimulus, *PLOS ONE*, 9:e109909, DOI: 10.1371/journal.pone.0109909.

13. Guterstam, A., Petkova, V. I., and Ehrsson, H. H. (2011), The illusion of owning a third arm, *PLOS ONE*, 6:e17208, DOI: 10.1371/journal.pone.0017208; Guterstam, A., Gentile, G., and Ehrsson, H. H. (2013), The invisible hand illusion: Multisensory integration leads to the embodiment of a discrete volume of empty space, *Journal of Cognitive Neuroscience*, 25:1078–1099.

14. Shokur, S., O'Doherty, J. E., Winans, J. A., Bleuler, H., Lebedev, M. A., and Nicolelis, M. A. L. (2013), Expanding the primate body schema in sensorimotor cortex by virtual touches of an avatar, *Proceedings of the National Academy of Sciences*, 110:15121–15126; Fang, W., Li, J., Qi, G., Li, S., Sigman, M., and Wang, L. (2019), Statistical inference of body representation in the macaque brain, *Proceedings of the National Academy of Sciences*, 116:20151–20157; Wada, M., Takano, K., Ora, H., Ide, M., and Kansaku, K. (2016), The rubber tail illusion as evidence of body ownership in mice, *Journal of Neuroscience*, 36:11133–11137.

15. Ehrsson, H. H., Spence, C., and Passingham, R. E. (2004), That's my hand! Activity in premotor cortex reflects feeling of ownership of a limb, *Science*, 305:875–877;

Karabanov, A. N., Ritterband-Rosenbaum, A., Christensen, M. S., Siebner, H. R., and Nielsen, J. B. (2017), Modulation of fronto-parietal connections during the rubber hand illusion, *European Journal of Neuroscience*, 45:964–974.

16. della Galla, F., Garbarini, F., Puglisi, G., Leonetti, A., Berti, A., and Borroni, P. (2016), Decreased motor cortex excitability mirrors own hand disembodiment during the rubber hand illusion, *eLife*, 5:e14972, DOI: 10.7554/eLife.14972.

17. Shokur, S., O'Doherty, J. E., Winans, J. A., Bleuler, H., Lebedev, M. A., and Nicolelis, M. A. L. (2013), Expanding the primate body schema in sensorimotor cortex by virtual touches of an avatar, *Proceedings of the National Academy of Sciences*, 110:15121–15126.

18. Azry, S., Overney, L. S., Landis, T., and Blanke, O. (2006), Neural mechanisms of embodiment: Asomatognosia due to premotor cortex damage, *Archives of Neurology* 63:1022–1025.

19. Gerstmann, J. (1942), Problem of imperception of disease and of impaired body territories with organic lesions, *Archives of Neurology & Psychiatry*, 48:890–913.

20. Dyer, C. (2000), Surgeon amputated healthy legs, *British Medical Journal*, 320:332.

21. Elliott, C. (2000), A new way to be mad, *Atlantic*, December.

22. Dua, A. (2010), Apotemnophilia: Ethical considerations of amputating a healthy limb, *Journal of Medical Ethics*, 36:75–78.

Chapter 4

1. Feinberg, T. E., and Farah, M. J. (2005), A historical perspective on cognitive neuroscience, in Farah, M. J., and Feinberg, M. E., eds., *Patient-Based Approaches to Cognitive Neuroscience*, Cambridge, MA: MIT Press, pp. 3–20; for discussion of three studies, see pp. 6–9.

2. Haggard, P., and Wolpert, D. M. (2005), Disorders of body scheme, in Freund, H.-J., Jeannerod, M., Hallet, M., and Leiguarda, R., eds., *Higher-Order Motor Disturbances*, Oxford: Oxford University Press; Goldenberg, G. (2000), Disorders of body perception, in Farah, M. J., and Feinberg, T. E., eds. *Patient-Based Approaches to Cognitive Neuroscience*, Cambridge, MA: MIT Press.

3. Tiemersma, D. (1989), *Body Image and Body Schema: An Interdisciplinary and Philosophical Study*, Amsterdam: Swets & Zeitlinger.

4. Head, H., and Holmes, G. (1911), Sensory disturbances from cerebral lesions, *Brain*, 34:102–254, 187.

5. Jacyna, L. S. (2016), *Medicine and Modernism: A Biography of Henry Head*, Pittsburgh: University of Pittsburgh Press.

6. Rivers, W. H. R., and Head, H. (1908), A human experiment in nerve division, *Brain*, 31:323–450.

7. Schilder, P. (1935/1978), *The Image and Appearance of the Human Body*, New York: International Universities Press.

8. Adler, A. (1965), The work of Paul Schilder, *Bulletin of the New York Academy of Medicine*, 41:842–53.

9. Critchley, M. (1950), The body-image in neurology, *Lancet*, 255:335–341.

10. Fisher, S., and Cleveland, S. E. (1956), Body-image boundaries and style of life, *Journal of Abnormal and Social Psychology*, 52:373–379; Fisher, S., and Fisher, R. L. (1964), Body image boundaries and patterns of body perception, *Journal of Abnormal Psychology*, 68:255–262.

11. Weckowicz, T. E., and Sommer, R. (1960), Body image and self-concept in schizophrenia. An experimental study, *Journal of Mental Science*, 106:17–39; Lukianowicz, L. (1967), "Body image" disturbances in psychiatric disorders, *British Journal of Psychiatry*, 113:31–47.

12. Cash, T. F. (2004), Body image: Past, present, and future, *Body Image*, 1:1–5.

13. York, G. K., III, and Steinberg, D. A. (2011), Hughlings Jackson's neurological ideas, *Brain*, 134:3106–3113.

14. Finger, S. (1994), *Origins of Neuroscience: A History of Explorations into Brain Function*, Oxford: Oxford University Press.

15. Penfield, W., and Boldrey, E. (1937), Somatic motor and sensory representation in the cerebral cortex of man as studied by electrical stimulation, *Brain*, 60:389–443.

16. Penfield, W., and Rasmussen, T. (1950), *The Cerebral Cortex of Man: A Clinical Investigation of the Localization of Function*, New York: Macmillan.

17. Schott, G. D. (1993), Penfield's homunculus: A note on cerebral cartography, *Journal of Neurology, Neurosurgery & Psychiatry*, 56:329–333.

18. Woolsey, T. A., and Van der Loos, H. (1970), The structural organization of layer IV in the somatosensory region (SI) of mouse cerebral cortex: The description of a cortical field composed of discrete cytoarchitectonic units, *Brain Research*, 17:205–242; Calford, M. B., Graydon, M. L., Huerta, M. F., Kaas, J. H., and Pettigrew, J. D. (1985), A variant of the mammalian somatotopic map in a bat, *Nature*, 313:477–479; Catania, K. C. (2011), The sense of touch in the star-nosed mole: From mechanoreceptors to brain, *Philosophical Transactions of the Royal Society B*, 366:3016–3025.

19. Kuehn, E., Dinse, J., Jakobsen, E., Long, X., Schäfer, A., Villringer, A., et al. (2017), Body topography parcellates human sensory and motor cortex, *Cerebral Cortex*, 27:3790–3805.

20. Di Noto, P. M., Newman, L., Wall, S., and Einstein, G. (2012), The *her*munculus: What is known about the representation of the female body in the brain?, *Cerebral Cortex*, 23:1005–1013.

21. Ramachandran, V. S., and McGeoch, P. D. (2008), Phantom penises in transsexuals, *Journal of Consciousness Studies*, 1:5–16; Case, L. K., Brang, D., Landazuri, R., Viswanathan, P., and Ramachandran, V. S. (2017), Altered white matter and sensory response to bodily sensation in female-to-male transgender individuals, *Archives of Sexual Behavior*, 46:1223–1227.

22. Ramón y Cajal, S. (1959), *Degeneration and Regeneration of the Nervous System*, New York: Hafner, p. 750 (original work published in 1913).

23. Merzenich, M. M., Nelson, R. J., Stryker, M. P., Cynader, M. S., Schoppmann, A., and Zook, J. M. (1983), Somatosensory cortical map changes following digit amputation in adult monkeys, *Journal of Comparative Neurology*, 224:591–605.

24. Clark, S. A., Allard, T., Jenkins, W. M., and Merzenich, M. M. (1988), Receptive fields in the body-surface map in adult cortex defined by temporally correlated inputs, *Nature*, 332:444–445.

25. Jenkins, W. M., Merzenich, M. M., Ochs, M. T., Allard, T., and Guic-Robles, E. (1990), Functional reorganization of primary somatosensory cortex in adult owl monkeys after behaviorally controlled tactile stimulation, *Journal of Neurophysiology*, 63:82–104.

26. Coq, J.-Q., and Xerri, C. (1999), Tactile impoverishment and sensorimotor restriction deteriorate the forepaw cutaneous map in the primary somatosensory cortex of adult rats, *Experimental Brain Research*, 129:518–531; Merzenich, M. M., Kaas, J. H., Wall, J., Nelson, R. J., Sur, M., and Felleman, D. (1983), Topographic organization of somatosensory cortical areas 3b and 1 in adult monkeys following restricted deafferentiation, *Neuroscience*, 8:33–55.

27. Elbert, T., Pantev, C., Wienbruch, C., Rockstroh, B., and Taub, E. (1995), Increased cortical representation of the fingers of the left hand in string players, *Science*, 270:305–307.

28. Lissek, S., Wilimzig, C., Stude, P., Pleger, B., Kalisch, T., Maier, C., et al. (2009), Immobilization impairs tactile perception and shrinks somatosensory cortical maps, *Current Biology*, 19:837–842.

29. Kolasinski, J., Makin, T. R., Logan, J. P., Jbabdi, S., Clare, S., Stagg, C. J., and Johansen-Berg, H. (2016), Perceptually relevant remapping of human somatotopy in 24 hours, *eLife*, 5:e17280.

30. Hahamy, A., Macdonald, S. N., van den Helligenberg, F., Kieliba, P., Emir, U., Malach, R., et al. (2017), Representation of multiple body parts in the missing-hand territory of congenital one-handers, *Current Biology*, 27:1–5; Dempsey-Jones,

H., Wesselink, D. B., Friedman, J., and Makin, T. R. (2019), Organized toe maps in extreme foot users, *Cell Reports*, 28:2748–2756.

31. Flor, H., Elbert, T., Knecht, S., Wienbruch, C., Pantev, C., Birbaumer, N., et al. (1995), Phantom-limb pain as a perceptual correlate of cortical reorganization following arm amputation, *Nature*, 375:482–484.

32. Ramachandran, V. S., and Rogers-Ramachandran, D. (2000), Phantom limbs and neural plasticity, *Archives of Neurology*, 57:317–320.

33. Clarke, S., Ragli, L., Janzer, R .C., Assal, G., and de Tribolet, N. (1994), Phantom face: Conscious correlates of neural reorganization after removal of primary sensory neurons, *NeuroReport*, 7:2853–2857.

34. Makin, T. R., Scholz, J., Filippini, N., Slater, D. H., Tracey, I., and Johansen-Berg, H. (2013), Phantom pain is associated with preserved structure and function in the former hand area, *Nature Communications*, 4:1570.

35. Makin, T. R., Scholz, J., Slater, D. H., Johansen-Berg, H., and Tracey, I. (2015), Reassessing cortical reorganization in the primary sensorimotor cortex following arm amputation, *Brain*, 138:2140–2146; Kikkert, S., Kolasinski, J., Jbabdi, S., Tracey, I., Beckmann, C. F., Johansen-Berg, H., and Makin, T. R. (2016), Revealing the neural fingerprints of a missing hand, *eLife*, 5:e15292.

36. Simões, E. L., Bramati, I., Rodrigues, E., Franzoi, A., Moll, J., Lent, R., and Tovar-Moll, F. (2012), Functional expansion of sensorimotor representation and structural reorganization of callosal connections in lower limb amputees, *Journal of Neuroscience*, 32:3211–3220.

37. Ramachandran, V. S., and Blakeslee, S. (2005), *Phantoms in the Brain*, London: Harper Perennial; McGeoch, P. D. (2007), Does cortical reorganisation explain the enduring popularity of foot-binding in medieval China?, *Medical Hypotheses*, 69:938–941.

38. Khateb, A., Simon, S. R., Dieguez, S., Lazeyras, F., Momjian-Mayor, I., Blanke, O., et al. (2009), Seeing the phantom: A functional magnetic resonance imaging study of a supernumerary phantom limb, *Annals of Neurology*, 65:698–705.

39. Medina, J., and Coslett, H. B. (2010), From maps to form to space: Touch and the body schema, *Neuropsychologia*, 48:645–654.

40. Longo, M. R., and Haggard, P. (2010), An implicit representation underlying human position sense, *Proceedings of the National Academy of Sciences*, 107:11727–11732; Longo, M. R., and Haggard, P. (2011), Weber's illusion and body shape: Anisotropy of tactile size perception on the hand, *Journal of Experimental Psychology: Human Perception & Performance*, 37:720–726; Linkenauger, S. A., Wong, H. Y., Guess, M., Stefanucci, J. K., McCulloch, K. C., Bülthoff, H. H., et al. (2015), The perceptual

homunculus: The perception of the relative proportions of the human body, *Journal of Experimental Psychology: General*, 144:103–113.

41. Brecht, M. (2017), The body model theory of somatosensory cortex, *Neuron*, 94:985–992.

42. Orlov, T., Makin, T. R., and Zohary, E. (2010), Topographic representation of the human body in the occipitotemporal cortex, *Neuron*, 68:586–600; Bracci, S., Caramazza, A., and Peelen, M. V. (2015), Representational similarity of body parts in human occipitotemporal cortex, *Journal of Neuroscience*, 35:12977–12985; Kikuchi, M., Takahashi, T., Hirosawa, T., Oboshi, Y., Yoshikawa, E., Minabe, Y., and Ouchi, Y. (2017), The lateral occipito-temporal cortex is involved in the mental manipulation of body part imagery, *Frontiers in Human Neuroscience*, 11:181, DOI: 10.3389/fnhum.2017.00181.

43. Weinstein, S., Sersen, E. A., and Vetter, R. J. (1964), Phantoms and somatic sensations in cases of congenital aplasia, *Cortex*, 1:276–290; Poeck, K. (1964), Phantoms following amputation in early childhood and in congenital absence of limbs, *Cortex*, 1:269–275.

44. Melzack, R., Israel, R., Lacroix, R., and Schultz, G. (1997), Phantom limbs in people with congenital limb deficiency or amputation in early childhood, *Brain*, 120:1603–1620; Tsakiris, M. (2010), *My* body in the brain: A neurocognitive model of body ownership, *Neuropsychologia*, 48:703–712.

Chapter 5

1. Ehrsson, H. H. (2007), The experimental induction of out-of-body experiences, *Science*, 317:1048; Lenggenhager, B., Tadi, T., Metzinger, T., and Blanke, O. (2007), Video ergo sum: Manipulating bodily self-consciousness, *Science*, 317:1096–1099; Guterstam, A., and Ehrsson, H. H. (2012), Disowning one's seen real body during an out-of-body illusion, *Consciousiousness & Cognition*, 21:1037–1042.

2. Petkova, V. I., and Ehrsson, H. H. (2008), If I were you: Perceptual illusion of body swapping, *PLOS ONE*, 3:e3832.

3. van der Hoort, B., Guterstam, A., and Ehrsson, H. H. (2011), Being Barbie: The size of one's own body determines the perceived size of the world, *PLOS ONE*, 6:e20195.

4. Kilteni, K., Normand, J.-M., Sanchez-Vivez, M. V., and Slater, M. (2012), Extending body space in immersive virtual reality: A very long arm illusion, *PLOS ONE*, 7:e40867.

5. Slater, M., Perez-Marcos, D., Ehrsson, H. H., and Sanchez-Vives, M. V. (2009), Inducing illusory ownership of a virtual body, *Frontiers in Neuroscience*, 3:214–220; Slater, M., Spanlang, B., Sanchez-Vives, M. V., and Blanke, O. (2010), First person experience of body transfer in virtual reality, *PLOS ONE*, 5:e10564.

6. Maister, L., Sebanz, N., Knoblich, G., and Tsakiris, M. (2016), Experiencing ownership over a dark-skinned body reduces implicit racial bias, *Cognition*, 128:170–178; Banakou, D., Hanumanthu, P. D., and Slater, M. (2016), Virtual embodiment of White people in a Black virtual body leads to a sustained reduction in their implicit racial bias, *Frontiers in Human Neuroscience*, 10:601; Tacikowski, P., Weijs, M. L., and Ehrsson, H. H. (2020), Perception of our own body influences self-concept and self-incoherence impairs episodic memory, *iScience*, https://doi.org/10.1016/j.isci.2020.101429.

7. Bréchet, L., Mange, R., Herbelin, B., Theillaud, Q., Gauthier, B., Serino, A., and Blanke, O. (2019), First-person view of one's body in immersive virtual reality: Influence on episodic memory, *PLOS ONE*, 14:e0197763.

8. Nishio, S., Watanabe, T., Ogawa, K., and Ishiguro, H. (2012), Body ownership transfer to teleoperated android robot, *Social Robotics*, 7621:398–407; Alimardini, M., Nishio, S., and Ishiguro, H. (2013), Humanlike robot hands controlled by brain activity arouse illusion of ownership in operators, *Scientific Reports*, 3:2396.

9. Ramachandran, V. S., and Altschuler, E. L. (2009), The use of visual feedback, in particular mirror visual feedback, in restoring brain function, *Brain*, 132:1693–1710.

10. Murray, C. D., Patchick, E., Pettifer, S., Howard, T., Caillette, F., Kulkarni, J., and Bamford, C. (2006), Investigating the efficacy of a virtual mirror box in treating phantom limb pain in a sample of chronic sufferers, *International Journal of Disability and Human Develepment*, 5:227–234.

11. Yu, X., Xie, Z., Yu, Y., Lee, J., Vasquez-Guardado, A., Luan, H., et al. (2019), Skin-integrated wireless haptic interfaces for virtual and augmented reality, *Nature*, 575:473–479.

12. Craske, B. (1977), Perception of impossible limb positions induced by tendon vibration, *Science*, 196:71–73; Jones, L. A. (1988), Motor illusions: What do they reveal about proprioception?, *Psychological Bulletin*, 103:72–86.

13. Lackner, J. R. (1988), Some proprioceptive influences on the perceptual representation of body shape and orientation, *Brain*, 111:281–297; Lackner, J. R., and DiZio, P. (2001), Somatosensory and proprioceptive contributions to body orientation, sensory localization, and self-calibration, in Nelson, R. J., ed., *The Somatosensory System: Deciphering the Brain's Own Body Image*, Boca Raton, FL: CRC Press, pp. 121–140.

14. Bell, C. (1834), *The Hand; Its Mechanism and Vital Endowments, as Evincing Design*, London: W. Pickering, p. 348.

15. Tuthill, J. C., and Azim, E. (2018), Proprioception, *Current Biology*, 28:R194–R203.

16. Mickle, A. D., Shepherd, A. J., and Mohapatra, D. P. (2017), Sensory TRP channels: The key transducers of nociception and pain, *Progress in Molecular Biology and Translational Sciience*, 131:73–118.

17. Chesler, A. T., Szczot, M., Bharucha-Goebel, D., Čeko, M., Donkervoort, S., Laubacher, C., et al. (2016), The role of PIEZO2 in human mechanosensation, *New England Journal of Medicine*, 375:1355–1364.

18. Cole, J. (1995), *Pride and a Daily Marathon*, Cambridge, MA: MIT Press, p. 12.

19. Guo, Y. R., and MacKinnon, R. (2017), Structure-based membrane dome mechanism for Piezo mechanosensitivity, *eLife*, 6:e33660.

20. Jousmäki, V., and Hari, R. (1998), Parchment-skin illusion: Sound-biased touch, *Current Biology*, 8:R190; Senna, I., Maravita, A., Bolognini, N., and Parise, C. V. (2014), The marble-hand illusion, *PLOS One*, 9:e91688.

21. Tajadura-Jiménez, A., Vakali, M., Fairhurst, M. T., Mandrigin, A., Bianchi-Berthouze, N., and Deroy, O. (2017), Contingent sounds change the mental representation of one's finger length, *Scientific Reports*, 8:4875; Tajadura-Jiménez, A., Deroy, O., Marquardt, T., Bianchi-Berthouze, N., Asai, T., Kimura, T., Kitagawa, N. (2018), Audio-tactile cues from an object's fall change estimates of one's body height, *PLoS ONE*, 13:e0199354.

22. Tajadura-Jiménez, A., Basia, M., Deroy, O., Fairhust, M., Marquardt, N., and Bianchi-Berthouze, N. (2015), As light as your footsteps: Altering walking sounds to change perceived body weight, emotional state and gait, in *Proceedings of the Thirty-third Annual ACM Conference on Human Factors in Computing Systems*, Seoul: ACM Press, 2943–2952.

23. Blanke, O. (2012), Multisensory brain mechanisms of bodily self-consciousness, *Nature Reviews Neuroscience*, 13:556–571.

24. Cowie, D., Makin, T. R., and Bremner, A. J. (2013), Children's responses to the rubber-hand illusion reveal dissociable pathways in body representation, *Psychological Science*, 24:762–769; Cowie, D., Sterling, S., and Bremner, A. J. (2016), The development of multisensory body representation and awareness continues to 10 years of age: Evidence from the rubber hand illusion, *Journal of Experimental Child Psychology*, 142:230–238.

25. Bufalari, I., Aprile, T., Avenanti, A., Di Russo, F., and Aglioti, S. M. (2007), Empathy for pain and touch in the human somatosensory cortex, *Cerebral Cortex*, 17:2553–2561; Filippetti, M. L., Lloyd-Fox, S., Longo, M. R., Farroni, T., Johnson, M. H. (2014), Neural mechanisms of body awareness in infants, *Cerebral Cortex*, 25:3779–3787; Saby, J. N., Meltzoff, A. N., and Marshall, P. J. (2015), Neural body maps in human infants: Somatotopic responses to tactile stimulation in 7-month-olds, *NeuroImage*, 118:74–78; Marshall, P. J., and Meltzoff, A. N. (2015), Body maps in the infant brain, *Trends in Cognitive Science*, 9:499–505; Boehme, R., Hauser, S., Gerling, G. J., Heilig, M., and Olausson, H. (2019), Distinction of self-produced touch and social touch at cortical and spinal cord levels, *Proceedings of the National Academy of Sciences*, 116:2290–2299.

26. Hilti, L. M., Hänggi, J., Vitacco, D. A., Kraemer, B., Palla, A., Luechinger, R., et al. (2013), The desire for healthy limb amputation: Structural brain correlates and clinical features of xenomelia, *Brain*, 136:318–329.

27. McGeoch, P. D., Brang, D., Song, T., Lee, R. R., Huang, M., and Ramachandran, V. S. (2011), Xenomelia: A new right parietal lobe syndrome, *Journal of Neurology, Neurosurgery & Psychiatry*, 82:1314–1319; Saetta, G., Hänggi, J., Gandola, M., Zapparoli, L., Salvato, G., Berlingeri, M., et al. (2020), Neural correlates of body integrity dysphoria, *Current Biology*, 30:1–5.

28. Wiesel, T. N., and Hubel, D. H. (1963), Single-cell responses in striate cortex of kittens deprived of vision in one eye, *Journal of Neurophysiology*, 26:1003–1017; Wiesel, T. N., and Hubel, D. H. (1965), Extent of recovery from the effects of visual deprivation in kittens, *Journal of Neurophysiology*, 28:1060–1072.

29. Money, J., Jobaris, R., and Furth, G. (1977), Apotemnophilia: Two cases of self-demand amputation as a paraphilia, *Journal of Sex Research*, 13:115–125.

30. Brugger, P., Christen, M., Jellstad, L., and Hänggi, J. (2016), Limb amputation and other disability desires as a medical condition, *Lancet Psychiatry*, 3:1176–1186.

31. Schilder, P. (1935/1978), *The Image and Appearance of the Human Body*, New York: International Universities Press, p. 216.

32. Blakemore, S.-J., Bristow, D., Bird, G., Frith, C., and Ward, J. (2005), Somatosensory activations during the observation of touch and a case of vision-touch synaesthesia, *Brain*, 128:1571–1583.

33. Ward, J., Schnakenberg, P., and Banissy, M. J. (2018), The relationship between mirror-touch synaesthesia and empathy: New evidence and a new screening tool, *Cognitive Neuropsychology*, 35:314–332.

Chapter 6

1. Wegner, D. M., Fuller, V. A., and Sparrow, B. (2003), Clever hands: Uncontrolled intelligence in facilitated communication, *Journal of Personality and Social Psychology*, 85:5–19.

2. Heyn, E. T. (1904), Berlin's wonderful horse: He can do almost anything but talk—How he was taught, *New York Times*, September 4, https://timesmachine.nytimes.com/timesmachine/1904/09/04/101396572.pdf.

3. Haggard, P., Clark, S., and Kalogeras, J. (2002), Voluntary action and conscious awareness, *Nature Neuroscience*, 5:382–385.

4. Caspar, E. A., Christensen, J. F., Cleeremans, A., and Haggard, P. (2015), Coercion changes the sense of agency in the human brain, *Current Biology*, 26:585–592.

5. Stetson, C., Cui, X., Montague, R., and Eagleman, D. M. (2006), Motor-sensory recalibration leads to an illusory reversal of action and sensation, *Neuron*, 51:651–659.

6. Della Sala, S. (2009), Dr. Strangelove syndrome, *Cortex*, 45:1278–1279.

7. Della Sala, S., Marchetti, C., and Spinnler, H. (1994), The anarchic hand: A fronto-mesial sign, in Boller, F., and Grafman, J., eds., *Handbook of Neuropsychology*, vol. 9:233–255, Amsterdam: Elsevier, as quoted in Della Sala, S. (2005), The anarchic hand, *Psychologist*, 18:606–609.

8. Della Sala, S., Marchetti, C., and Spinnler, H. (1991), Right-sided anarchic (alien) hand: A longitudinal study, *Neuropsychologia*, 11:1113–1127.

9. Amalnath, S. D., Subramanian, R., and Dutta, T. K. (2013), The alien hand sign, *Annals of the Indian Academy of Neurology*, 16:9–11.

10. Kornhuber, H. H., and Deecke, L. (1965), Hirnpotentialänderungen bei Willkürbewegungen und passiven Bewegungen des Menschen: Bereitschaftspotential und reafferente Potentiale, *Pflügers Archive für die Gesamte Physiologie*, 284:1–17; quotations from Deecke, L. (2014), Experiments into readiness for action: 50th anniversary of the Bereitschaftspotential, *World Neurology*, 29:1.

11. Libet, B., Gleason, C. A., Wright, E. W., and Pearl, D. K. (1983), Time of conscious intention to act in relation to onset of cerebral activity (readiness-potential): The unconscious initiation of a freely voluntary act, *Brain*, 106:623–642.

12. Soon, C. S., Brass, M., Heinze, H.-J., and Haynes, J.-D. (2008), Unconscious determinants of free decisions in the human brain, *Nature Neuroscience*, 11:543–545.

13. Fried, I., Mukamel, R., and Kreiman, G. (2011), Internally generated preactivation of single neurons in human medial frontal cortex predicts volition, *Neuron*, 69:548–562.

14. Shurger, A., Sitt, J. D., and Dehaene, S. (2012), An accumulator model for spontaneous neural activity prior to self-initiated movement, *Proceedings of the National Academy of Sciences*, 109:E2904–E2913.

15. John Hughlings Jackson, as quoted in Pearce, J. M. S. (2009), Hugo Karl Liepmann and apraxia, *Clinical Medicine*, 9:466–470.

16. Pearce, J. M. S. (2009), Hugo Karl Liepmann and apraxia, *Clinical Medicine*, 9:466–470.

17. Frith, C. D., Blakemore, S.-J., and Wolpert, D. M. (2000), Abnormalities in the awareness and control of action, *Philosophical Transactions of the Royal Society of London B*, 355:1771–1788; Blakemore, S. J., and Frith, C. D. (2003), Self-awareness and action, *Current Opinion in Neurobiology*, 13:219–224.

18. Blakemore, S.-J., and Frith, C. D. (2003), Self-awareness and action, *Current Opinion in Neurobiology,* 13:219–224; Thomas, A. K., Loftus, E. F. (2002), Creating bizarre false memories through imagination, *Memory and Cognition,* 30:423–431.

19. Desmurget, M., Reilly, K. T., Richard, N., Szathmari, A., Mottolese, C., and Sirigu, A. (2009), Movement intention after parietal cortex stimulation in humans, *Science,* 324:811–813; Danckert, J., Ferber, S., Doherty, T., Steinmetz, H., Nicolle, D., and Goodale, M. A. (2002), Selective, non-lateralized impairment of motor imagery following right parietal damage, *Neurocase,* 8:194–204.

20. Blakemore, S.-J., and Frith, C. D. (2003), Self-awareness and action, *Current Opinion in Neurobiology,* 13:219–224.

21. Maruff, P., Wilson, P., and Currie, J. (2003), Abnormalities of motor imagery associated with somatic passivity phenomena in schizophrenia, *Schizophrenia Research,* 60:229–238.

22. Synofzik, M., Thier, P., Leube, D. T., Schlotterbeck, P., and Linder, A. (2010), Misattributions of agency in schizophrenia are based on imprecise predictions about the sensory consequences of one's actions, *Brain,* 133:262–271.

23. Wenke, D., Fleming, S. M., and Haggard, P. (2010), Subliminal priming of actions influences sense of control over effects of action, *Cognition,* 115: 26–38.

24. Synofzik, M., Vosgerau, G., and Newen, A. (2008), Beyond the comparator model: A multifactorial two-step account of agency, *Consciousiousness and Cognition,* 17:219–239.

25. Frith, C. D., Blakemore, S.-J., and Wolpert, D. M. (2000), Abnormalities in the awareness and control of action, *Philosophical Transactions of the Royal Society of London B,* 355:1771–1788; Della Sala, S. (2005), The anarchic hand, *Psychologist,* 18:606–609.

26. Marchetti, C., and Della Sala, S. (1998), Disentangling the alien and anarchic hand, *Cognitive Neuropsychiatry,* 3:191–207.

27. *New York Times* (1904), "Clever Hans" again. Expert commission decides that the horse actually reasons, October 2, https://timesmachine.nytimes.com/timesmachine/1904/10/02/120289067.pdf.

28. Samhita, L., and Gross, H. J. (2013), The "Clever Hans phenomenon" revisited, *Communicative & Integrative Biology,* 6:e27122.

29. Carpenter, W. B. (1852), *On the Influence of Suggestion in Modifying and Directing Muscular Movement, Independently of Volition,* London: Royal Institution of Great Britain.

30. Noë, A. (2006), *Action in Perception,* Cambridge, MA: MIT Press, p. 1.

31. Held, R., and Hein, A. (1963), Movement-produced stimulation in the development of visually guided behavior, *Journal of Comparative Physiology and Psychology*, 56:872–876.

32. Witt, J. K. (2018), Perception and action, in J. T. Wixted, ed., *Stevens' Handbook of Experimental Psychology and Cognitive Neuroscience*, vol. 2: *Sensation, Perception, and Attention*, pp. 489–523, New York: Wiley.

Chapter 7

1. Carroll, L. (1866), *Alice's Adventures in Wonderland*, with forty-two illustrations by John Tenniel, London: Macmillan.

2. Fine, E. J. (2013), The Alice in Wonderland syndrome, *Progress in Brain Research*, 206:143–156; Farooq, O., and Fine, E. J. (2017), Alice in Wonderland syndrome: A historical and medical review, *Pediatric Neurology*, 77:5–11.

3. Lippman, C. W. (1952), Certain hallucinations peculiar to migraine, *Journal of Nervous & Mental Disorders*, 116:346–351.

4. Todd, J. (1955), The syndrome of Alice in Wonderland, *Canadian Medical Association Journal*, 73:701–704.

5. Blau, J. N. (1998), Somesthetic aura: The experience of "Alice in Wonderland," *Lancet*, 352:582; Podoll, K., and Robinson, D. (1999), Lewis Carroll's migraine experiences, *Lancet*, 353:1366.

6. Fine, E. J. (2013), The Alice in Wonderland syndrome, *Progress in Brain Research*, 206:143–156; Blom, J. D. (2016), Alice in Wonderland syndrome: A systematic review, *Neurology Clinical Practice*, 6:259–270; Beh, S. C., Masrour, S., Smith, S. V., and Friedman, D. I. (2018), Clinical characteristics of Alice in Wonderland syndrome in a cohort with vestibular migraine, *Neurology Clinical Practice*, 8:389–396.

7. Pearce, J. M. S. (2004), Richard Morton: Origins of Anorexia Nervosa, *European Neurology*, 52:191–192.

8. Niedzielski, A., Kaźmierczak, N., and Grzybowski, A. (2017), Sir William Withey Gull (1816–1890), *Journal of Neurology*, 264:419–420.

9. Vandereycken, W., and van Deth, R. (1990), A tribute to Lasègue's description of anorexia nervosa (1873), with completion of its English translation, *British Journal of Psychiatry*, 157:902–908.

10. Bruch, H. (1962), Perceptual and conceptual disturbances in anorexia nervosa, *Psychosomatic Medicine*, 24:187–194.

11. Slade, P. D., and Russell, G. F. M. (1973), Awareness of body dimensions in anorexia nervosa: Cross-sectional and longitudinal studies, *Psychological Medicine*, 3:188–199.

12. Dillon, D. J. (1962), Measurement of perceived body size, *Perceptual and Motor Skills*, 14:191–196.

13. Allebeck, P., Hallberg, D., and Espmark, S. (1976), Body image—an apparatus for measuring disturbances in estimation of size and shape, *Journal of Psychosomatic Research*, 20:583–589.

14. Traub, A. C., and Orbach, J. (1964), Psychophysical studies of body-image: 1. The adjustable body-distorting mirror, *Archives of General Psychiatry*, 11:53–66.

15. Caspar, R. C., Halmi, K. A., Goldberg, S. C., Eckert, E. D., and Davis, J. M. (1979), Disturbances of body image estimation as related to other characteristics and outcome in anorexia nervosa, *British Journal of Psychiatry*, 134:60–66.

16. Lask, B., Gordon, I., Christie, D., Frampton, I., Chowdury, U., and Watkins, B. (2005), *International Journal of Eating Disorders*, 37:S49–S51; Suchan, B., Busch, M., Schulte, D., Grönermeyer, D., and Vocks, S. (2009), Reduction of gray matter density in the extrastriate body area in women with anorexia nervosa, *Behavioural Brain Research*, 206:63–67.

17. Keiser, A., Smeets, M. A. M., Dijkerman, H. C., van den Hout, M., Klugkist, I., van Elburg, A., and Postma, A. (2011), Tactile body image disturbance in anorexia nervosa, *Psychiatry Research*, 190:115–120.

18. Keiser, A., Smeets, M. A. M., Dijkerman, H. C., Uzunbajakau, S. A., van Elburg, A., and Postma, A. (2013), Too fat to fit through the door: First evidence for disturbed body-scaled action in anorexia nervosa during locomotion, *PLOS ONE*, 8:e64602.

19. Piryankova, I. V., Wong, H. Y., Linkenauger, S. A., Stinson, C., Longo, M. R., Bülthoff, H. H., and Mohler, B. J. (2014), Owning an overweight or underweight body: Distinguishing the physical, experienced and virtual body, *PLOS ONE*, 9:e103428.

20. Keiser, A., van Elburg, A., Helms, R., and Dijkerman, H. C. (2016), A virtual reality full body illusion improves body image disturbance in anorexia nervosa, *PLOS ONE*, 11:e0163921.

21. Hardy, G. E., and Cotterill, J. A. (1982), A study of depression and obsessionality in dysmorphophobic and psoriatic patients, *British Journal of Psychiatry*, 140:19–22.

22. Krebs, G., de la Cruz, L. F., and Mataiz-Cols, D. (2017), Recent advances in understanding and managing body dysmorphic disorder, *Evidence Based Mental Health*, 20:71–75.

23. Phillips, K. A. (2005), *The Broken Mirror: Understanding and Treating Body Dysmorphic Disorder*, Oxford: Oxford University Press, p. 20.

24. Pope, H. G., Jr., Gruber, A. J., Choi, P., Olivardia, R., and Phillips, K. A. (1997), Muscle dysmorphia: An underrecognized form of body dysmorphic disorder, *Psychosomatics*, 38:548–557.

25. Grace, S. A., Labuschagne, I., Kaplan, R. A., and Rossell, S. L. (2017), The neurobiology of body dysmorphic disorder: A systematic review and theoretical model, *Neuroscience and Biobehavioural Reviews*, 83:83–96.

26. Gilpin, H. R., Moseley, G. L., Stanton, T. R., and Newport, R. (2015), Evidence for distorted mental representation of the hand in osteoarthritis, *Rheumatology*, 54:678–682; Moseley, G. L. (2005), Distorted body image in complex regional pain syndrome, *Neurology*, 65:773.

27. Schwoebel, J., Friedman, R., Duda, N., and Coslett, H. B. (2001), Pain and the body schema: Evidence for peripheral effects on mental representations of movement, *Brain*, 124:2098–2104.

28. Maihöfner, C., Handwerker, H. O., Neundörfer, B., and Birklein, F. (2003), Patterns of cortical reorganization in complex regional pain syndrome, *Neurology*, 61:1707–1715; Mancini, F., Wang, A. P., Schira, M. M., Isherwood, Z. J., McAuley, J. H., Ianetti, G. D., et al. (2019), Fine-grained mapping of cortical somatotopies in chronic complex regional pain syndrome, *Journal of Neuroscience*, 39:9185–9196.

29. Dhond, R. P., Ruzich, E., Witzel, T., Maeda, Y., Malatesta, C., Morse, L. R., et al. (2012), Spatio-temporal mapping cortical neuroplasticity in carpal tunnel syndrome, *Brain*, 135:3062–3073.

30. Maeda, Y., Kettner, N., Holden, J., Lee, J., Kim, J., Cina, S., et al. (2014), Functional deficits in carpal tunnel syndrome reflect reorganization of primary somatosensory cortex, *Brain*, 137:1741–1752.

31. Gandevia, S. C., and Phegan, C. M. L. (1999), Perceptual distortions of the human body image produced by local anaesthesia, pain and cutaneous stimulation, *Journal of Physiology*, 514:609–616.

32. Paqueron, X., Leguen, M., Rosenthal, D., Coriat, P., Willer, J. C., and Danziger, N. (2003), The phenomenology of body image distortions induced by regional anaesthesia, *Brain*, 126:702–712; Paqueron, X., Gentili, M. E., Willer, J. C., Coriat, P., and Riou, B. (2004), Time sequence of sensory changes after upper extremity block: Swelling sensation is an early and accurate predictor of success, *Anesthesiology*, 101:162–168.

33. Silva, S., Loubinoux, I., Olivier, M., Bataille, B., Fourcade, O., Samii, K., et al. (2011), Impaired visual hand recognition in preoperative patients during brachial plexus anesthesia: Importance of peripheral neural input for mental representation of the hand, *Anesthesiology*, 114:126–134.

34. Moseley, G. L., Parsons, T. J., and Spence, C. (2008), Visual distortion of a limb modulates the pain and swelling evoked by movement, *Current Biology*, 18: R1047–R1048.

35. Preston, C., and Newport, R. (2011), Analgesic effects of multisensory illusions in osteoarthritis, *Rheumatology*, 50:2314–2315.

36. Longo, M. R., Betti, V., Aglioti, S. M., and Haggard, P. (2009), Visually induced analgesia: Seeing the body reduces pain, *Journal of Neuroscience*, 29:12125–12130.

37. Longo, M. R., Ianetti, G. D., Mancini, F., Driver, J., and Haggard, P. (2012), Linking pain and the body: Neural correlates of visually induced analgesia, *Journal of Neuroscience*, 32:2601–2607.

38. Alfred Serko, as quoted in Critchley, M. (1950), The body-image in neurology, *Lancet*, 255:335–341.

39. Takaoka, K., and Takata, T. (1999), "Alice in Wonderland" syndrome and lilliputian hallucinations in a patient with a substance-related disorder, *Psychopathology*, 32:47–49.

Chapter 8

1. Heathers, J. A. J., Fayn, K., Silvia, P. J., Tiliopoulos, N., and Goodwin, M. S. (2018), The voluntary control of piloerection, *PeerJ*, 6:e5292; DOI: 10.7717/peerj.5292.

2. Terem, I., Ni, W. W., Goubran, M., Rahimi, M. S., Zaharchuk, G., Yeom, K. W., Moseley, M. E., et al. (2018), Revealing sub-voxel motions of brain tissue using phase-based amplified MRI (aMRI), *Magnetic Resonance in Medicine*, 80:2549–2559; Mosher, C. P., Wei, Y., Kamiński, J., Nandi, A., Mamelak, A. N., Anastassiou, C. A., and Rutihauser, U. (2020), Cellular classes in the human brain revealed *in vivo* by heartbeat-related modulation of the extracellular action potential waveform, *Cell Reports*, 30:3536–3551.

3. Garfinkel, S. N., Barrett, A. B., Minati, L., Dolan, R. J., Seth, A. K., and Critchley, H. D. (2013), What the heart forgets: Cardiac timing influences memory for words and is modulated by metacognition and interoceptive sensitivity, *Psychophysiology*, 50:505–512.

4. Azevedo, R. T., Garfinkel, S. N., Critchley, H. D., and Tsakiris, T. (2017), Cardiac afferent activity modulates the expression of racial stereotypes, *Nature Communications*, 8:13854.

5. Ohl, S., Wohltat, C., Kliegl, R., Pollatos, O., and Engbert, R. (2016), Microsaccades are coupled to heartbeat, *Journal of Neuroscience*, 36:1237–1241.

6. Zelano, C., Jiang, H., Zhou, G., Arora, N., Schuele, S., Rosenow, J., and Gottfried, J. A. (2016), Nasal respiration entrains human limbic oscillations and modulates cognitive function, *Journal of Neuroscience*, 36:12448–12467.

7. Tsakiris, M., Tajadura-Jiménez, A., and Costantini, M. (2011), Just a heartbeat away from one's body: Interoceptive sensitivity predicts malleability of body-representations, *Proceedings of the Royal Society B*, 278:2470–2476; Suzuki, K., Garfinkel, S. N., Critchley, H. D., and Seth, A. K. (2013), Multisensory integration across exteroceptive interoceptive domains modulates self-experience in the rubber-hand

illusion, *Neuropsychologia*, 51:2909–2917; Aspell, J. E., Heydrich, L., Marillier, G., Lavanchy, T., Herbelin, B., and Blanke, O. (2013), Turning body and self inside out: Visualized heartbeats alter bodily self-consciousness and tactile perception, *Psychological Science*, 24:2445–2453.

8. Heydrich, L., Aspell, J. E., Marillier, G., Lavanchy, T., Herbelin, B., and Blanke, O. (2018), Cardio-visual full body illusion alters bodily self-consciousness and tactile processing in somatosensory cortex, *Scientific Reports*, 8:9230; Al, E., Iliopoulos, F., Forschack, N., Nierhaus, T., Grund, M., Motyka, P., et al. (2020), Heart-brain interactions shape somatosensory perception and evoked potentials, *Proceedings of the National Academy of Sciences*, 117:10575–10584; Motyka, P., Grund, M., Forschack, N., Al, E., Villringer, A., and Gaebler, M. (2019), Interactions between cardiac activity and conscious somatosensory perception, *Psychophysiology*, 56:e13424.

9. Herman, A. M., and Tsakiris, M. (2020), Feeling in control: The role of cardiac timing in the sense of agency, *Affective Science*, 1:155–171, https://doi.org/10.1007/s42761-020-00013-x.

10. Craig, A. D. (2003), Interoception: The sense of the physiological condition of the body, *Current Opinion in Neurobiology*, 13:500–505; Craig, A. D. (2009), How do you feel—now? The anterior insula and human awareness, *Nature Reviews Neuroscience*, 10:59–70.

11. Critchley, H. D., Mathias, C. J., Josephs, O., O'Doherty, J., Zanini, S., Dewar, B. K., et al. (2003), Human cingulate cortex and autonomic control: Converging neuroimaging and clinical evidence, *Brain*, 126:2139–2152; Salomon, R., Ronchi, R., Dönz, J., Bello-Ruiz, J., Herbelin, B., Martet, R., et al. (2016), The insula mediates access to awareness of visual stimuli presented synchronously to the heartbeat, *Journal of Neuroscience*, 36:5115–5127.

12. Park, H.-D., and Blanke, O. (2019), Coupling inner and outer body for self-consciousness, *Trends in Cognitive Science*, 23:377–388.

13. Nummenmaa, L., Hari, R., Hietanen, J. K., and Glerean, E. (2018), Maps of subjective feelings, *Proceedings of the National Academy of Sciences*, 115:9198–9203.

14. James, W. (1884), What is an emotion?, *Mind*, 9:188–205; main-text James quotations on pp. 189, 190, 189–190, and 190; emphasis in original.

15. Seth, A. K., and Friston, K. J. (2016), Active interoceptive inference and the emotional brain, *Philosophical Transactions of the Royal Society B*, 371:20160007; Barrett, L. F., Quigley, K. S., and Hamilton, P. (2016), An active inference theory of allostasis and interoception in depression, *Philosophical Transactions of the Royal Society B*, 371:20160011.

16. Pollatos, O., Herbert, B. M., Matthias, E., and Schandry, R. (2007), Heart rate response after emotional picture presentation is modulated by interoceptive awareness, *International Journal of Psychophysiology*, 63:117–124; Jung, W.-M., Ryu, Y., Lee,

Y.-S., Wallraven, C., and Chae, Y. (2017), Role of interoceptive accuracy in topographical changes in emotion-induced bodily sensations, *PLOS ONE*, 12:e0183211.

17. Ainley, V., Apps, M. A. J., Fotopoulou, A., and Tsakiris, M. (2016), "Bodily precision": A predictive coding account of individual differences in interoceptive accuracy, *Philosophical Transactions of the Royal Society B*, 371:20160003:

18. Brewer, R., Cook, R., and Bird, G. (2016), Alexithymia: A general deficit of interoception, *Royal Society Open Science*, 3:150664.

19. Bruch, H. (1962), Perceptual and conceptual disturbances in anorexia nervosa, *Psychsomatic Medicine*, 24:187–194.

20. Fassino, S., Pierò, A., Gramaglia, C., and Abbate-Daga, G. (2004), Clinical, psychopathological and personality correlates of interoceptive awareness in anorexia nervosa, bulimia nervosa and obesity, *Psychopathology*, 37:168–174; Pollatos, O., Kurz, A.-L., Albrecht, J., Schreder, T., Kleeman, A. M., Schöpf, V., et al. (2008), Reduced perception of bodily signals in anorexia nervosa, *Eating Behavior*, 9:381–388; Khalsa, S. S., Graske, M. G., Li, W., Vangala, S., Strober, M., and Feussner, J. D. (2015), Altered interoceptive awareness in anorexia nervosa: Effects of meal anticipation, consumption and bodily arousal, *International Journal of Eating Disorders*, 48:889–897.

21. Oberndorfer, T., Simmons, A., McCurdy, D., Strigo, I., Matthews, S., Yang, T., et al. (2013), Greater anterior insula activation during anticipation of food images in women recovered from anorexia nervosa versus controls, *Psychiatry Research*, 30:132–141; Wagner, A., Aizenstein, H., Mazurkewicz, L., Fudge, J., Frank, G. K., Putnam, K., et al. (2008), Altered insula response to taste stimuli in individuals recovered from restricting-type anorexia nervosa, *Neuropsychopharmacology*, 33:513–523; Kerr, K. L., Moseman, S. E., Avery, J. A., Bodurka, J., Zucker, N. L., and Simmons, W. K. (2016), Altered insula activity during visceral interoception in weight-restored patients with anorexia nervosa, *Neuropsychopharmacology*, 41:521–528.

22. Paulus, M. P., and Stein, M. B. (2010), Interoception in anxiety and depression, *Brain Structure and Function*, 214:451–463; Hielscher, E., Whitford, T. J., Scott, J. G., and Zopf, R. (2019), When the body is the target—Representations of one's own body and bodily sensations in self-harm, *Neuroscience and Bevahiour Reviews*, 101:85–112; DeVille, D. C., Kuplicki, R., Stewart, J. L., Tulsa 1000 Investigators, Paulus, M. P., and Khalsa, S. S. (2020), Diminished responses to bodily threat and blunted interoception in suicide attempters, *eLife*, 9:51593.

Chapter 9

1. Hall, E. T. (1966), *The Hidden Dimension,* New York: Anchor Books, pp. 119–125.

2. Hall, *The Hidden Dimension*, p. x.

3. Hall, *The Hidden Dimension*, p. 63.

4. Rizzolatti, G., Scandolara, C., Gentilucci, M., and Camarda, R. (1981), Response properties and behavioral modulation of "mouth" neurons of the postarcuate cortex (area 6) in macaque monkeys, *Brain Research*, 255:421–424.

5. Graziano, M. S. A., Yap, G. S., and Gross, C. G. (1994), Coding of visual space by premotor neurons, *Science*, 266:1054–1057; Graziano, M. A. S., Hu, X. T., and Gross, C. G. (1997), Coding the locations of objects in the dark, *Science*, 277:239–241.

6. Graziano, M. S. A. (2016), Ethological action maps: A paradigm shift for the motor cortex, *Trends in Cognitive Science*, 20:121–132.

7. Brain, W. R. (1941), Visual disorientation with special reference to lesions of the right cerebral hemisphere, *Brain*, 64:244–272.

8. Bjoertomt, O., Cowey, A., and Walsh, V. (2002), Spatial neglect in near and far space investigated by repetitive transcranial magnetic stimulation, *Brain*, 125: 2012–2022.

9. Longo, M. R., and Lourenco, S. F. (2007), Space perception and body morphology: Extent of near space scales with arm length, *Experimental Brain Research*, 177:285–290.

10. Lourenco, S. F., Longo, M. R., and Pathman, T. (2011), Near space and its relation to claustrophobic fear, *Cognition*, 119:448–453.

11. Ferri, F., Tajadura-Jiménez, A., Väljamäe, A., Vastano, R., and Costantini, M. (2015), Emotion-inducing approaching sounds shape the boundaries of multisensory peripersonal space, *Neuropsychologia*, 70:468–475.

12. Serino, A., Noel, J.-P., Galli, G., Canzoneri, E., Marmaroli, P., Lissek, H., and Blanke, O. (2015), Body part–centered and full body–centered peripersonal space representations, *Scientific Reports*, 22:18603.

13. Bufacchi, R. J., and Iannetti, G. D. (2018), An action field theory of peripersonal space, *Trends in Cognitive Science*, 22:1076–1090.

14. Striem-Amit, E., Vannuscorps, G., and Caramazza, A. (2017), Sensorimotor-independent development of hands and tools selectivity in the visual cortex, *Proceedings of the National Academy of Sciences*, 114:4787–4792.

15. Iriki, A., Tanaka, M., and Iwamura, Y. (1996), Coding of modified body schema during tool use by macaque postcentral neurons, *NeuroReport*, 7:2325–2330.

16. Farnè, A., and Làdavas, E. (2000), Dynamic size-change of hand peripersonal space following tool use, *NeuroReport*, 11:1645–1649; Farnè, A., Iriki, A., and Làdavas, E. (2005), Shaping multisensory action–space with tools: Evidence from patients with cross-modal extinction, *Neuropsychologia*, 43:238–248.

17. Head, H., and Holmes, G. (1911), Sensory disturbances from cerebral lesions, *Brain*, 34:102–254.

18. Critchley, M. (1950), The body-image in neurology, *Lancet*, 225:335–341.

19. Cardinali, L., Frassinetti, F., Brozzoli, C., Urquizar, C., Roy, A. C., and Farnè, A. (2009), Tool-use induces morphological updating of the body schema, *Current Biology*, 19:R478–R479.

20. Miller, L. E., Montroni, L., Koun, E., Salemme, R., Hayward, V., and Farnè, A. (2018), Sensing with tools extends somatosensory processing beyond the body, *Nature*, 561:239–242.

21. Miller, L. E., Fabio, C., Ravenda, V., Bahmad, S., Koun, E., Salemme, R., et al. (2019), Somatosensory cortex efficiently processes touch located beyond the body, *Current Biology*, 29:1–8.

22. Quallo, M. M., Price, C. J., Ueno, K., Azamizuya, T., Cheng, K., Lemon, R. N., and Iriki, A. (2009), Gray and white matter changes associated with tool-use learning in macaque monkeys, *Proceedings of the National Academy of Sciences*, 106:18379–18384.

23. Miller, L. E., Longo, M. R., and Saygin, A. P. (2017), Visual illusion of tool use recalibrates tactile perception, *Cognition*, 162:32–40.

24. Baccarini, M., Martel, M., Cardinali, L., Sillan, O., Farnè, A., and Roy, A. C. (2014), Tool use imagery schema triggers tool incorporation in the body schema, *Frontiers in Psychology*, 5:492.

25. Baccarini et al., Tool use imagery schema.

26. Cardinali, L., Brozzoli, C., and Farnè, A. (2009), Peripersonal space and body schema: Two labels for the same concept?, *Brain Topology*, 21:252–260.

27. Nerlich, A. G., Zink, A., Szeimies, U., and Hagedorn, H. G. (2000), Ancient Egyptian prosthesis of the big toe, *Lancet*, 356:2176–219; Finch, J. L. (2011), The ancient origins of prosthetic medicine, *Lancet*, 377:548–549; Finch, J. L., Heath, G. H., David, A. R., and Kulkamin, J. (2012), Biomechanical assessment of two artificial big toe restorations from Ancient Egypt and their significance to the history of prosthetics, *Journal of Prosthetics and Orthotics*, 24:181–191.

28. Hernigou, P. (2013), Ambroise Paré IV: The early history of artificial limbs (from robotic to prostheses), *International Orthopaedics*, 37:1195–1197.

29. Warwick, K., Gasson, M., Hutt, B., Goodhew, I., Kyberd, P., Andrews, B., et al. (2003), The application of implant technology for cybernetic systems, *Archives of Neurology*, 60:1369–1373.

30. van den Heiligenberg, F. M. Z., Orlov, T., Macdonald, S. N., Duff, E. P., Slater, D. H., Beckmann, C. F., et al. (2018), Artificial limb representation in amputees, *Brain*, 141:1422–1433.

31. Kuiken, T. A., Miller, L. A., Lipschutz, R. D., Lock, B. A., Stubblefield, K., Marasco, P. D., and Dumanian, G. A. (2007), Targeted reinnervation for enhanced prosthetic arm function in a woman with a proximal amputation: A case study, *Lancet*, 369:371–380.

32. Marasco, P. D., Kim, K., Colgate, J. E., Peshkin, M. A., and Kuiken, T. A. (2011), Robotic touch shifts perception to a prosthesis in targeted reinnervation amputees, *Brain*, 134:747–758.

33. Clites, T., Carty, M. J., Ullauri, J. B., Carney, M. E., Mooney, L. M., Duval, J.-M., Srinivasan, S. S., and Herr, H. M. (2018), Proprioception from a neurally controlled lower-extremity prothesis, *Science Translational Medicine*, 10:eaap8373.

34. Marasco, P. D., Hebert, J. S., Sensinger, J. W., Shell, C. E., Schofield, J. S., Thumser, Z. C., et al. (2018), Illusory movement perception improves motor control for prosthetic hands, *Science Translational Medicine*, 10:eaao6990.

35. D'Anna, E., Valle, G., Mazzoni, A., Strauss, I., Iberite, F., Patton, J., et al. (2019), A closed-loop hand prosthesis with simultaneous intraneural tactile and position feedback, *Science Robotics*, 4:eaau8892.

36. Petrini, F. M., Bumbasirevic, M., Valle, G., Ilic, V., Mijović, P., Čvančara, P., Barberi, F., Katic, N., Bortolotti, D., Andreu, D., Lechler, K., Lesic, A., Mazic, S., Mijović, B., Guiraud, D., Stieglitz, T., Alexandersson, A., Micera, S., and Raspopovic, S., et al. (2019), Sensory feedback restoration in leg amputees improves walking speed, metabolic cost and phantom pain, *Nature Medicine*, 25:1356–1363.

37. Rognini, G., Petrini, F. M., Raspopovic, S., Valle, G., Granata, G., Strauss, I., et al. (2019), Multisensory bionic limb to achieve prosthesis embodiment and reduce distorted phantom limb perceptions, *Journal of Neurology, Neurosurgery, and Psychiatry*, 90:833–836.

38. Oddo, C. M., Mazzoni, A., Spanne, A., Enander, J. M. D., Mogensen, H., Bengtsson, F., et al. (2017), Artificial spatiotemporal touch inputs reveal complementary decoding in neocortical neurons, *Scientific Reports*, 7:45898; Osborn, L. E., Dragomir, A., Betthauser, J. L., Hunt, C. L., Nguyen, H. H., Kaliki, R. R., and Thaker, N. V. (2018), Prosthesis with neuromorphic multilayered e-dermis perceives touch and pain, *Science Robotics*, 3:eaat3818; Zhao, H., O'Brien, K., Li, S., and Shepherd, R. F. (2016), Optoelectronically innervated soft prosthetic hand via stretchable optical waveguides, *Science Robotics*, 1:eaai7529; Lee, W. W., Tan, Y. J., Yao, H., Li, S., See, H. H., Hon, M., et al. (2019), A neuro-inspired artificial peripheral nervous system for scalable electronic skins, *Science Robotics*, 4:eaax2198; Sonar, H. A., Gerratt, A. P., Lacour, S. P., and Paik, J. (2019), Closed-loop haptic feedback control using a self-sensing soft pheumatic actuator skin, *Soft Robotics*, 7:22–29.

39. Ortiz-Catalan, M., Mastinu, E., Sassu, P., Aszmann, O., and Brånemark, R. (2020), Self-contained neuromusculoskeletal arm prostheses, *New England Journal of Medicine*, 382:1732–1738.

Chapter 10

1. Farahany, N. A., Greely, H. T., Hyman, S., Koch, C., Grady, C., Paşca, S. P., et al. (2018), The ethics of experimenting with human brain tissue, *Nature*, 556:429–432.

2. Clayton, N. S., Dally, J. M., and Emery, N. J. (2007), Social cognition by food-caching corvids: The western scrub-jay as a natural psychologist, *Philosophical Transactions of the Royal Society of London B*, 362:507–522; Raby, C. R., Alexis, D. M., Dickinson, A., and Clayton, N. S. (2007), Planning for the future by western scrub-jays, *Nature*, 445:919–921; Gruber, R., Schiestl, M., Boeckle, M., Frohnwieser, A., Miller, R., Gray, R. D., et al. (2019), New Caledonian crows use mental representations to solve metatool problems, *Current Biology*, 29:686–697.e3.

3. Brooks, W. R. (1989), Hermit crabs alter sea anemone placement patterns for shell balance and predation, *Journal of Experimental Marine Biology and Ecology*, 132:109–112; Sonoda, K., Asakura, A., Minoura, M., Elwood, R. W., and Gunji, Y. P. (2012), Hermit crabs perceive the extent of their virtual bodies, *Biology Letters*, 23:495–497.

4. Ravi, S., Siesenop, T., Bertrand, O., Li, L., Doussot, C., Warren, W. H., Combes, S. A., and Engelhaaf, M. (2020), Bumblebees perceive the spatial layout of their environment in relation to their body size and form to minimize inflight collisions, *Proceedings of the National Academy of Sciences*, 117:31494–31499.

5. Krause, T., Spindler, L., Poeck, B., and Strauss, R. (2019), *Drosophila* acquires a long-lasting body-size memory from visual feedback, *Current Biology*, 29:1833–1841.

6. Bongard, J., Zykov, V., and Lipson, H. (2006), Resilient machines through continuous self-modeling, *Science*, 314:1118–1121; Kwiatkowski, R., and Lipson, H. (2019), Task-agnostic self-modeling machines, *Science Robotics*, 4:eaau9354.

Index